Graphene Photonics, Optoelectronics, and Plasmonics

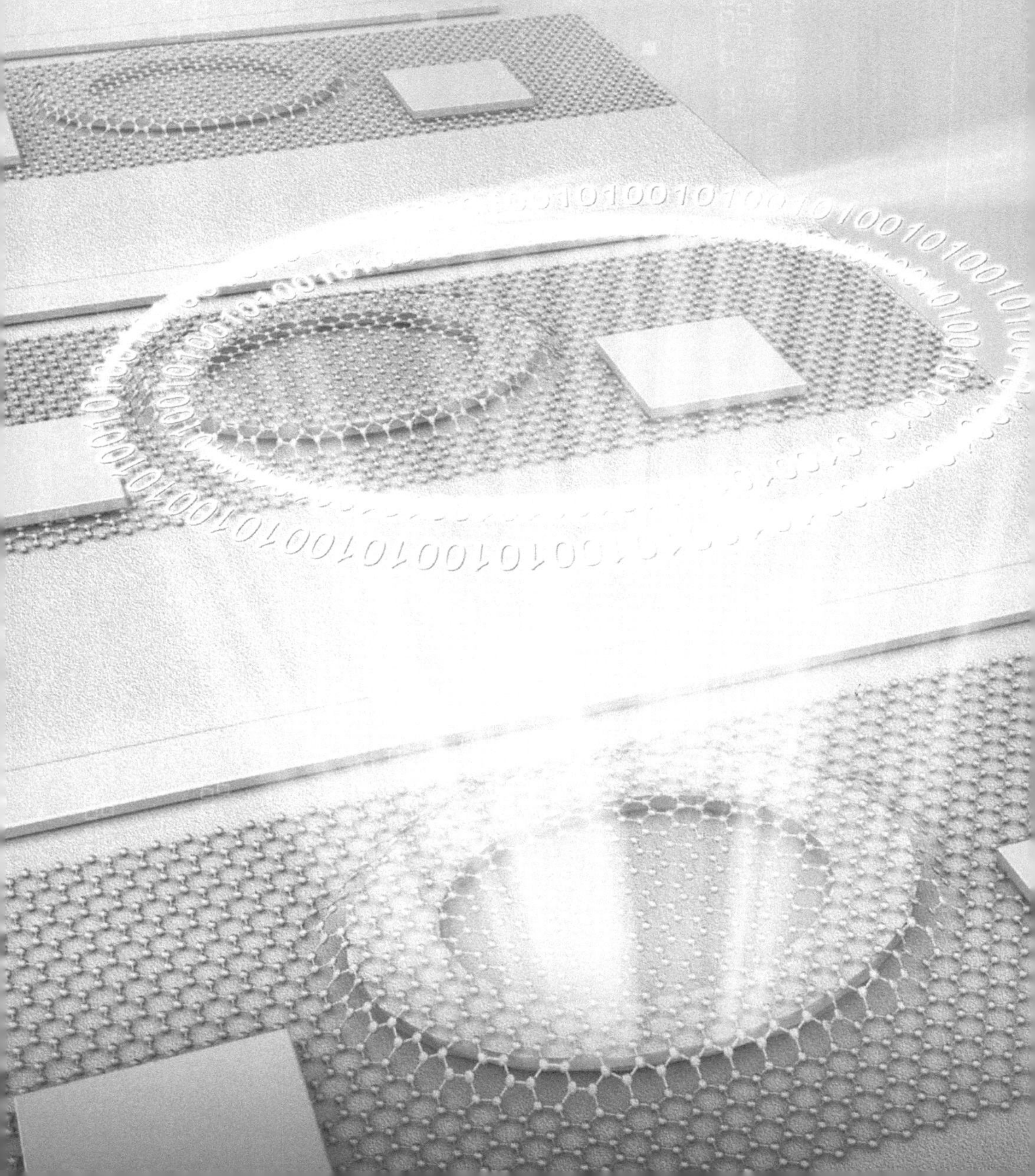

Graphene Photonics, Optoelectronics, and Plasmonics

edited by

Qiaoliang Bao
Hui Ying Hoh
Yupeng Zhang

PAN STANFORD PUBLISHING

Published by

Pan Stanford Publishing Pte. Ltd.
Penthouse Level, Suntec Tower 3
8 Temasek Boulevard
Singapore 038988

Email: editorial@panstanford.com
Web: www.panstanford.com

British Library Cataloguing-in-Publication Data
A catalogue record for this book is available from the British Library.

Graphene Photonics, Optoelectronics, and Plasmonics

ISBN 978-981-4745-60-4 (Hardcover)
ISBN 978-1-315-19667-1 (eBook)

Printed in the USA

Contents

Preface xi

1 Introduction **1**
Shivananju Bannur Nanjunda, Hui Ying Hoh, Joice Sophia, Yunzhou Xue, Jian Yuan, Wenzhi Yu, Haoran Mu, Yao Chen, and Qiaoliang Bao
1.1 Introduction 1
1.2 Discovery of Graphene and Its Properties 3
1.3 Production of Graphene 5
1.3.1 Mechanical Exfoliation 6
1.3.2 Epitaxial Growth 8
1.3.2.1 Epitaxial growth mechanism 8
1.3.2.2 Different substrates adopted 8
1.3.3 Chemical Vapor Deposition 11
1.3.4 Solution Processing 14
1.3.4.1 Graphite/graphene oxides 14
1.3.4.2 Reduced graphene oxides 15
1.4 Optical Properties of Graphene 17
1.4.1 Linear Optical Properties 17
1.4.2 Nonlinear Optical Properties 19

2 Application of Graphene in Lasers **27**
Yao Chen, Haoran Mu, Yupeng Zhang, and Qiaoliang Bao
2.1 Introduction 27
2.2 Mode-Locking Fiber Laser Based on Graphene 29
2.3 Q-Switching Fiber Laser Based on Graphene 32
2.4 Graphene in the Solid State Laser 33
2.5 Conclusion 37

3 Graphene-Based Optical Modulators **41**
Sheng Gan, Yupeng Zhang, and Qiaoliang Bao
3.1 Introduction 41
3.2 Graphene Strip-Waveguide Optical Modulators 42
3.3 Graphene-Slot Waveguide Optical Modulators 44
3.4 Optical Modulators with MZI and Ring Structures 47
3.5 Optical Modulators with Vertical Structure 50
3.6 Conclusion 53

4 Graphene-Based Optical Polarizers **57**
Ziyu Wang, Zai-Quan Xu, Yupeng Zhang, and Qiaoliang Bao
4.1 Introduction 57
4.2 Conventional In-Line Polarizers 58
4.3 Graphene-Based Polarizers 59
4.3.1 Optical Fiber Polarizers 59
4.3.2 Optical Waveguide Polarizers 61
4.4 Conclusion 62

5 Graphene-Based Photodetectors **65**
Shaojuan Li, Yan Liu, Hui Ying Hoh, and Qiaoliang Bao
5.1 Introduction 65
5.2 Plasma-Enhanced Graphene Photodetectors 69
5.3 Resonant Cavity–Enhanced Graphene Photodetectors 70
5.4 Waveguide-Integrated Graphene Photodetector 72
5.5 Van der Waals Heterojunction Structure Photodetector 74
5.6 Conclusion and Outlook 76

6 Graphene for Solar Cells **81**
Pengfei Li, Yusheng Wang, Qasim Khan, Hui Ying Hoh, and Qiaoliang Bao
6.1 Introduction 81
6.2 Different Types of Graphene for Solar Cells 82
6.3 Dye-Sensitized Solar Cells 83
6.3.1 Graphene as Transparent Electrodes 83
6.3.2 Graphene as Catalytic Counter Electrodes 85
6.4 Organic/Polymer Solar Cells 87
6.4.1 Functionalized Graphene as Electrode 87

6.4.2 Functionalized Graphene as Transport Layer 88
6.4.3 Functionalized Graphene as Active Layer 90
6.5 Graphene-Silicon Hybrid Solar Cells 94
6.6 Conclusion 96

7 Graphene Plasmonics 101
Qingyang Xu, Yao Lu, Jian Yuan, Yupeng Zhang, and Qiaoliang Bao
7.1 Introduction 101
7.2 Theoretical Investigations 102
7.2.1 Electron Behavior in Two Dimensions 105
7.2.2 Dispersion Relation of Graphene Plasmons 105
7.2.3 Surface Plasmon Polaritons 107
7.2.4 Surface Plasmons Coupled with Phonons 108
7.3 Excitation of Surface Plasmons in Graphene 110
7.3.1 Spectroscopy Investigations of Surface Plasmons 110
7.3.2 Real-Space Imaging and Detection of Surface Plasmons 112
7.3.2.1 EELS and ARPES 113
7.3.2.2 Light–plasmons coupling 114
7.4 Plasmonic Devices 120
7.4.1 Surface Plasmons in Graphene 121
7.4.2 Surface Plasmons in Graphene Ribbons 122
7.4.3 Surface Plasmons in Graphene Disks, Rings 122
7.4.4 Surface Plasmons in Graphene/Insulator or TI Stacks 123
7.5 Conclusion 124

8 Graphene as Optical Limiters 131
Hui Ying Hoh and Qiaoliang Bao
8.1 Introduction 131
8.2 Nonlinear Effect 133
8.2.1 Nonlinear Absorption—Multiphoton Absorption 133
8.2.2 Nonlinear Absorption—Excited-State Absorption (Reverse Saturable Absorption) 133
8.2.3 Nonlinear Scattering 134

8.3 Carbon-Based Optical Limiters 134
8.4 Graphene-Based Optical Limiters 136
8.4.1 Graphene Dispersions 136
8.4.2 Composites Containing Graphene 142
8.5 Conclusion 143

9 Graphene-Based Light-Emitting Diodes 147
Jialu Zheng, Hui Ying Hoh, Yupeng Zhang, and Qiaoliang Bao
9.1 Introduction 147
9.2 Development of Graphene-Based LEDs 148
9.2.1 Selective Doped Graphene as Anodic Electrode or Cathodic Electrode 148
9.2.2 Improvement in CVD Growth of Graphene 151
9.3 Combination of Graphene Electrodes with Other Materials 153
9.3.1 Integration of Graphene Electrodes with Nanoclusters 153
9.3.2 Integration of Graphene Electrodes with One-Dimensional Semiconductor Nanostructures 154
9.3.3 Integration of Graphene Electrodes with SWCNTs 155
9.3.4 Integration of Graphene Electrodes with Nanowires 156
9.4 Other Potential Applications of Graphene in LEDs 157
9.4.1 Photoluminescence from Graphene Oxide 157
9.4.2 Graphene Interlayer for Current Spreading Enhancement 158
9.5 Conclusion 158

10 Graphene-Based Touchscreens 163
Shivananju Bannur Nanjunda and Qiaoliang Bao
10.1 Introduction 163
10.2 Transparent Touchscreen Electrodes 164
10.2.1 Indium Tin Oxide 165
10.2.2 Conducting Polymers 165
10.2.3 Carbon Nanotubes 166

10.2.4 Graphene 166
10.2.5 Working Principle of Graphene Touchscreen 167
10.3 Resistive Touchscreens 167
10.4 Capactive Touchscreens 170
10.5 Graphene-Based Touchscreens 172
10.5.1 Comparison of Graphene and ITO-Based Touchscreen Devices 173
10.5.2 Current and Future Trends of Graphene-Based Touchscreens 174

11 Perspective **177**
Hui Ying Hoh and Qiaoliang Bao
11.1 Strategies for Overcoming Challenges 179
11.2 Other Two-Dimensional Materials 180

Index 187

Preface

We have been investigating the optical applications of graphene since 2008 because we saw that it has the potential to significantly improve human lives in several ways such as energy harvesting (solar cells), optical communication, and low-cost touchscreens. There is numerous literature about graphene synthesis and properties and some on devices and applications. However, only few provide in-depth discussions about the many possible ways graphene may be utilized in optical devices, and this gave us the idea to bring out *Graphene Photonics, Optoelectronics, and Plasmonics*.

Photonics is the science and technology concerned with light (photons or particles of light), and it deals with its properties, generation, manipulation and, detection. Optoelectronics refers to the electronic effects induced by light or devices relying on such working principles. Plasmonics, a relatively newer field, refers to the coupling of light to charges like free electrons in metals. These electrons can be excited by the electric component of light, resulting in collective oscillations (plasmons).

This book is organized on the basis of the different applications presented. This will allow readers, especially seasoned investigators, who have a practical problem in mind to jump to the relevant chapter quickly. While we have tried to include many applications, this book is not meant to be exhaustive and we encourage interested readers to peruse through other literature.

This book is also targeted toward tertiary students and researchers who are venturing into this field. The first chapter is an introductory chapter which discusses the discovery, processing methods, and optical properties of graphene and is especially useful for readers who are less acquainted with graphene. The last chapter highlights the latest developments and our views on advancements of graphene.

We thank the publisher for taking us on this journey. It has been extremely rewarding. Special thanks go out to Stanford Chong and the excellent editorial team, particularly Shivani Sharma.

We are grateful to our colleagues for their wonderful input, both directly as contributors in this book or indirectly through discussions and suggestions. It has been a great pleasure working with them, and we trust they feel the same, too.

Finally, we thank our family for their unwavering support.

Qiaoliang Bao
Hui Ying Hoh
Yupeng Zhang
Spring 2017

Chapter 1

Introduction

Shivananju Bannur Nanjunda,[a] Hui Ying Hoh,[b] Joice Sophia,[a] Yunzhou Xue,[a] Jian Yuan,[a] Wenzhi Yu,[a] Haoran Mu,[a] Yao Chen,[a] and Qiaoliang Bao[a,b]

[a] *Institute of Functional Nano & Soft Materials (FUNSOM), Soochow University, 199 Ren-Ai Road, Suzhou Industrial Park, Suzhou Jiangsu 215123, China*
[b] *Department of Materials Science and Engineering, Monash University, 22 Alliance Lane, Building 69, Clayton, 3800, Victoria, Australia*
qiaoliang.bao@gmail.com

1.1 Introduction

The Greek word "graphein" meaning "to write" has been well known to humans for centuries. In the 1940s, graphene had been discussed only theoretically, but it took almost 60 years to obtain a high-quality, two-dimensional one-atom-thick layer experimentally. In 2004, Andre Geim and Konstantin Novoselov at the University of Manchester successfully isolated graphene; for that, they were awarded the Nobel Prize in physics in 2010.

Graphene is a single sheet of sp^2-hybridized carbon atoms bonded in a honeycomb lattice, or more simply put "one single layer of the graphite structure" [1]. In retrospect, graphene can be viewed as the basic building block of all graphitic forms. Buckyballs,

Graphene Photonics, Optoelectronics, and Plasmonics
Edited by Qiaoliang Bao, Hui Ying Hoh, and Yupeng Zhang

ISBN 978-981-4745-60-4 (Hardcover), 978-1-315-19667-1 (eBook)
www.panstanford.com

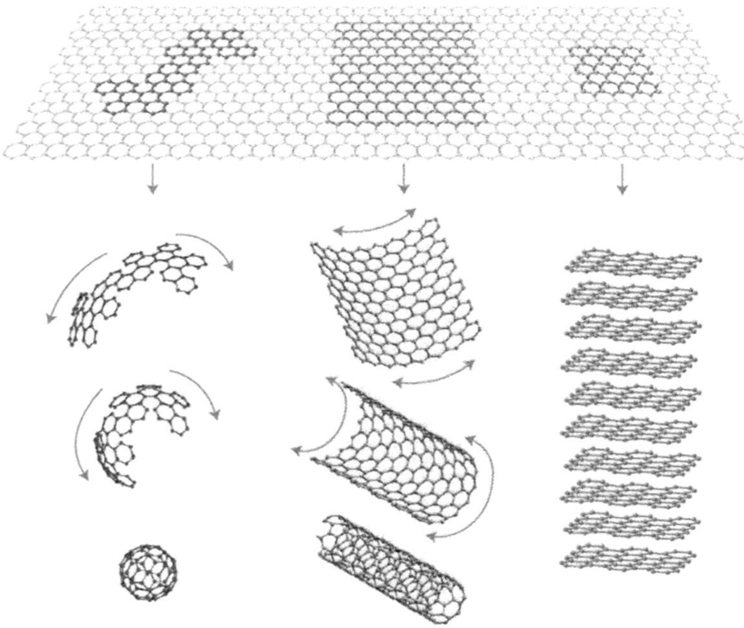

Figure 1.1 Schematic of graphene, depicting how buckyballs, nanotubes, and graphite can be formed from it. Reprinted by permission from Macmillan Publishers Ltd: *Nature Materials*, Ref. 1, Copyright 2007.

nanotubes, and graphite all originate from graphene, as depicted in Fig. 1.1 [1, 2].

Among this new class of materials, the first to be discovered is the zero-dimensional buckminsterfullerene, a carbon cluster of 60 carbon atoms, which forms a spherical cage looking like a soccer ball. Since its discovery in 1985, C_{60} and other fullerenes were found to be naturally occurring, e.g., in candle soot, initiating excitement in low-dimensional forms of carbon among the research community [3]. The one-dimensional carbon nanotubes were later discovered in 1991 by Iijima [4]. Nanotubes are, as described by their name, nano-sized cylinders formed by rolling up a graphene sheet. Finally, the last form of carbon that completes the series, successfully isolated only in 2004, is the two-dimensional graphene [5].

1.2 Discovery of Graphene and Its Properties

The discovery, or successful isolation, of graphene can be described as an intentional accident. Professor Geim from Manchester University wanted to study the properties of thin layers of graphite (10–100 layers) and assigned the task of obtaining such layers to a student. After some time, his student polished a block of graphite to a small grain, although a useless one as it was essentially still bulk graphite.

Transfixed with the initial objective and remembering how surfaces were cleaned by pealing the top layers with scotch tape, Geim and another student Novoselov started playing around with graphite flakes that were stuck on the tape. Eventually, it was found that thin flakes can be transferred to a silicon wafer (with 300 nm SiO_2) with the sticky tape, and that they are observable under an optical microscope (Fig. 1.2) [1, 2]. This finding is a breakthrough as it allows graphene of one, two, or more layers to be prepared and detected without sophisticated equipment. The isolated crystals, although small (less than 1 mm), are of high quality and perfect for fundamental experiments [1, 2].

The amazing mechanical, electrical, thermal, and optical properties of graphene are undeniable. Transport measurements revealed that graphene has a "mobility"—the speed at which an electrical charge flows across a semiconductor—of up to 250 times that of

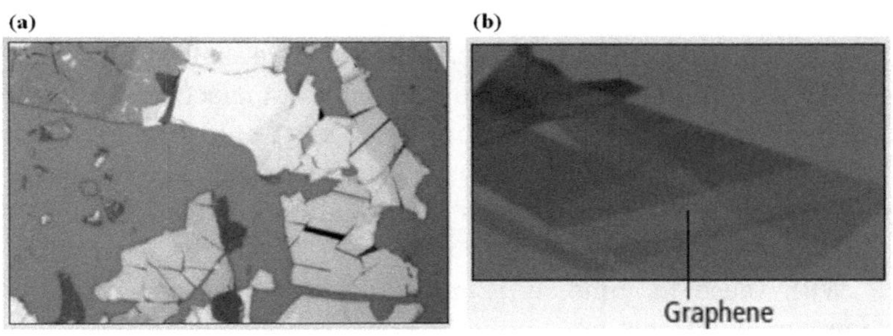

Figure 1.2 Left: Graphite flakes on an oxidized silicon surface. Right: Graphene, highly transparent, on a similar surface. Image courtesy: Professor Pablo Jarillo-Herrero, Department of Physics, Massachusetts Institute of Technology.

silicon. Carrier mobilities exceeding 200,000 $cm^2V^{-1}s^{-1}$ have been reported, and even more amazingly, the charge carriers in graphene propagate through the lattice with a much decreased effective mass, described by the Dirac equation instead of the Schrödinger equation [5–10].

The mechanical and physical properties are just as fascinating. Graphene is 1 million times thinner than a strand of human hair but three times stronger than steel. It is as pliable as rubber and can be stretched to 120% of its length. The Young's modulus is around 1100 GPa, while the fracture strength is 125 GPa. It is the most flexible and stretchable crystal, yet the most impermeable material, even to helium. Owing to strong covalent bonding with the lattice, graphene also has high thermal conductivity of about 5000 $Wm^{-1}K^{-1}$, several times that of copper [5–9].

Graphene also exhibits remarkable optical properties. For example, it can be optically visualized, despite being only a single atom thick. Its transmittance (T) can be expressed in terms of the fine-structure constant. The linear dispersion of the Dirac electrons makes broadband applications possible. Saturable absorption is observed as a consequence of Pauli blocking, and non-equilibrium carriers result in hot luminescence. Chemical and physical treatments can also lead to luminescence [10]. The optical properties of graphene will be further elaborated in Section 1.4.

The richness of optical and electronic properties of graphene renders it an ideal photonic and optoelectronic material and attracts enormous interest. So far, the main focus has been on fundamental physics and electronic devices. However, we believe the true potential of graphene lies in photonics and optoelectronics, where the combination of its unique optical and electronic properties can be fully exploited, even in the absence of a bandgap, and the linear dispersion of the Dirac electrons enables ultra-wideband tunability [10].

With these in mind, it is not surprising that research and commercial interest in graphene grew exponentially (Fig. 1.3). In recent years, studies on graphene are steadily diverging from fundamental investigations and steering toward practical functions, ranging from coatings, additives, energy storage, electronics, as well as optoelectronics and photonics applications. With its

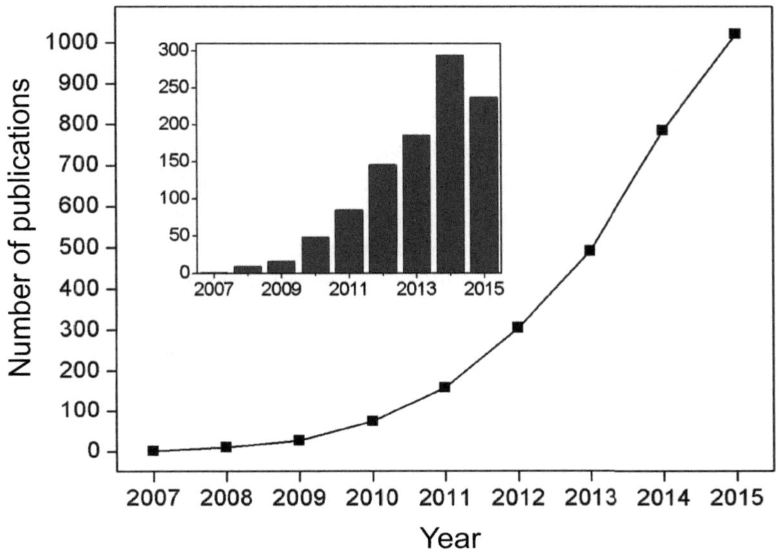

Figure 1.3 Number of reports (inset: patents) containing the term "graphene" by year, determined by searching for "graphene" in the Scopus database.

high chemical stability, mechanical strength, and carrier mobility, graphene is a promising candidate for future high-technology applications. Graphene-based photonic and optoelectronic devices such as photodetectors, ultrafast lasers, polarizers, modulators, light-emitting diodes, photovoltaic devices, touchscreens, transparent conductors, optical limiters, terahertz devices, optical frequency converters, flexible smart windows, and display systems are already in the production line. As graphene obtained from different methods have varying properties, it is critical to have a basic understanding of the production methods, so that one may make an informed choice of a suitable form of graphene for the intended application [10].

1.3 Production of Graphene

Graphene was isolated for the first time using the scotch tape technique (where thinner strips are peeled off a block of graphite using sticky tape), and the quantities we can make in large areas still lag behind this method. Exfoliation of bulk graphite (the top-down approach) is the most commonly used method for the mass production of small graphene sheets. This can be through direct exfoliation in a liquid, with or without the use of a surfactant, or in the solid state by edge functionalization, or by first inserting a chemical species between the graphene layers in graphite to weaken their interaction followed by exfoliation [1, 2].

Chemical assembly of small graphene sheets and chemical vapor deposition (CVD) are two methods to produce large-area graphene films. Chemical assembly can be easily realized on various substrates at low temperatures, but the films obtained suffer from poor quality, for example low electrical conductivity. CVD produces high-quality graphene films by the catalytic decomposition of hydrocarbons on a metal (e.g., Cu, Ni, Pt, or alloy) surface at high temperatures, and the films are then transferred to transparent substrates such as glass and polymers by etching away the metal or by non-destructive electrochemical bubbling for transparent conductive film (TCF) applications. There has been a lot of work to try and enhance the manufacturability of graphene and there are a few techniques that look very promising but are not completely mature yet [11–28].

Recently, Samsung electronics has demonstrated a single layer of graphene, which is 30 inches in diameter. So in just a few years, we have gone all the way from micrometer-sized flakes to 30 inches [23]. The ambition is that one day graphene will be fabricated in the same way as printing newspapers—in a roll-to-roll. This will change the entire economics of the optical and electronics industry.

1.3.1 Mechanical Exfoliation

Graphite consists of many layers of graphene (Fig. 1.4a). In order to exfoliate a single sheet, the weak van der Waals attraction between different layers must be overcome without disturbing any subsequent sheets. If we apply a small force to the graphite, like rubbing it, we can get the graphene easily (Fig. 1.4b). The original top-down approach of mechanical exfoliation produces high-quality samples, but the method is neither high throughput nor high yielding [1, 2].

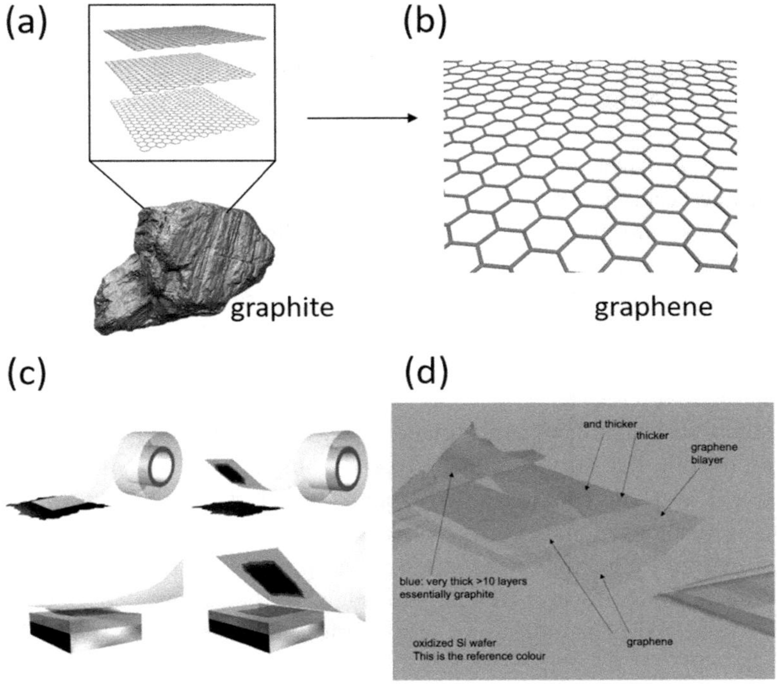

Figure 1.4 (a) Graphite, the common allotrope of carbon, consists of a stack of many graphene layers held together by weak (van der Waals) interactions. (b) Graphene is a sheet of carbon, one-atom thick, arranged in a honeycomb lattice. (c) Flow diagram of mechanical exfoliation. Figure reprinted from Ref. 56, with permission from IOP Publishing, Copyright 2012, The Royal Swedish Academy of Sciences. (d) Optical image of different layers of graphene. Image courtesy: Professor Pablo Jarillo-Herrero, Department of Physics, Massachusetts Institute of Technology.

The main process of mechanical exfoliation involves using the scotch tape to stick to both sides of the highly oriented pyrolytic graphite (HOPG) and repeating this process many times until we get a single layer of graphene (Fig. 1.4c). Thin films of graphene are transparent to the naked eye. However, on a silicon oxide substrate, the added graphene layers interfere with the light, causing a shift in colors, allowing one to distinguish between bilayer and multilayer graphene (Fig. 1.4d) [1, 2].

1.3.2 Epitaxial Growth

A highly motivating technique for the production of high-quality and large-scale graphene films is the epitaxial growth of graphene. The search for a suitable substrate has so far been a major issue in the epitaxial growth of graphene since it imposes serious constraints in device architecture and functionality. Hence, there is an increasing interest in finding specific dielectrics that allow substrate-supported geometry [11–19].

Ultrathin epitaxial graphene (EG) films grown on several substrates are promising candidates in the field of nanoelectronics. Growing large areas of EG on silicon carbide (SiC) is technological feasible; moreover, SiC, being a semiconductor, is a convenient substrate for subsequent stages of device fabrication. Due to its atomically flat surface almost free of dangling bonds and charge traps and its capability to design the electronic structure of graphene, hexagonal boron nitride (h-BN) is also an attractive substrate for nanodevices based on graphene [15–17]. The lattice constant of h-BN is relatively equivalent to that of graphite having wide electrical bandgap and large optical phonon modes (Fig. 1.5) [11–19].

1.3.2.1 Epitaxial growth mechanism

The epitaxial growth mechanisms of graphene at the atomic level and the effects of substrates by examining the transient stages at the growing edges of graphene have been well explained by Ozcelik et al. [11]. The formation of epitaxial graphene can be classified into two steps in which nucleation is the first step followed by graphene growth from the nucleated seed. The defects that form the step edges in atomic scale play a leading role in the process of nucleation at the substrate. The epitaxial growth of graphene on different substrates, such as Ru(0001), Pt(111), Ir(111), Cu(111), and Pd(111), was previously reported. Readers interested in the growth mechanism and factors influencing that are encouraged to refer to these studies [11–19].

1.3.2.2 Different substrates adopted

Graphene on SiC: Graphene grown on structured SiC surfaces have been found to overcome edge roughness, creating a great

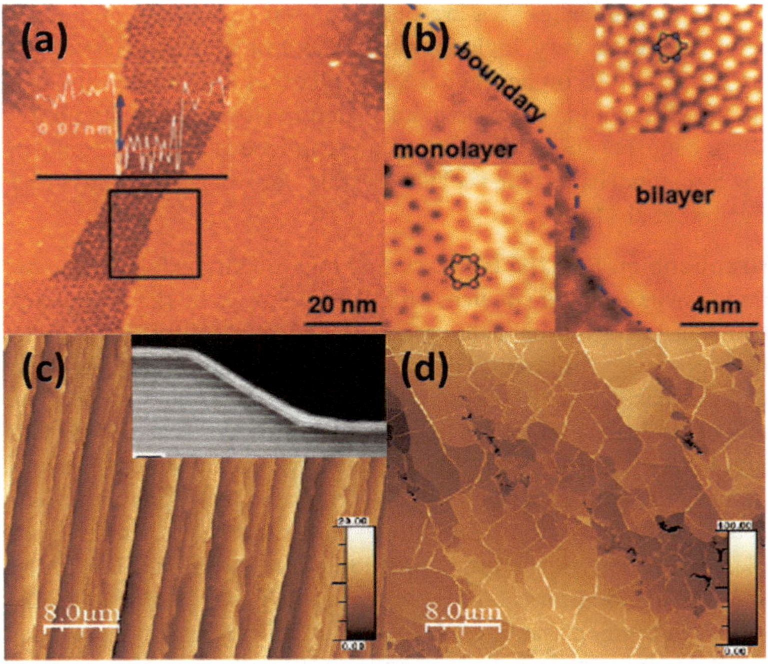

Figure 1.5 (a) Scanning tunnelling microscopy (STM) image of the monolayer (darker channel region in the center) and bilayer (two bright terraces in the right and left parts) EG on 6H-SiC (0001). (b) High-resolution STM image showing the coexistence of monolayer (lower left part) and bilayer (upper right part) EG. Figures (a) and (b) reprinted with permission from Ref. 12, Copyright 2008, American Chemical Society. (c) Atomic force micrograph (AFM) of EG grown on Si-face 6H-SiC with inset (left) as TEM cross section showing EG over a Si-face 6H-SiC step bunch and (d) AFM of EG grown on C-face 4H-SiC having a network of ridges ("giraffe stripes") can be seen (height scales are in nm). Figures (c) and (d) reproduced from Ref. 13 by permission of The Electrochemical Society.

opportunity for nanoscale patterning. EG grown on SiC can be patterned, which shows quantum confinement of electrons, two-dimensional electron gas behavior, and enhanced photosensitivity. On the other hand, the polar nature of 6H(4H)-SiC substrate leads to two non-equivalent surfaces of (0001), also called Si face, and (000-1), known as C face, in which the resulting multilayer graphene over these two surfaces shows different structures and properties. Consequently, samples grown on the C-terminated face exhibit electronic properties that are rather similar to those of monolayer graphene with very high mobility. The transport and structural properties of EG on hexagonal SiC depict that the charge carries in EG are chiral along with the band structure clearly correlated to the Dirac cone. For more information on graphene grown epitaxially on the hexagonal basal planes of SiC crystals, one can refer to the review by First et al. [14].

Graphene on BN: The fabrication of high-quality exfoliated mono- and bilayer graphene devices on h-BN substrates by the mechanical transfer process has been found to have mobility and carrier inhomogeneity approximately an order of magnitude superior to devices based on SiO_2 with reduced roughness, intrinsic doping, and chemical reactivity. The epitaxial growth of single-domain graphene on h-BN with a fixed stacking orientation by plasma-assisted deposition method has been reported. The epitaxial growth is attained by cracking the methane molecules by means of remote plasma source prior to the deposition of graphene on h-BN. In this way, the demand for catalysts, as seen in CVD, can be excluded. The main achievement has been the ability to seal monolayer and bilayer graphene in a fixed orientation on the h-BN substrate, as evidenced by the equivalent uniform trigonal moiré patterns repeated on all the graphene patches [15–17].

Graphene on sapphire by van der Waals epitaxy: The van der Waals epitaxy can be described as the epitaxial growth of a layered material, which proceeds with a van der Waals force owing to good epitaxy in spite of the large lattice mismatch between the grown and substrate materials. The epilayers start to grow with their own lattice constants in the beginning but have very few defects with the interface formed due to the weak van der Waals interaction. The lattice mismatch of the individual materials involved in the epitaxy is a major challenge in obtaining a good epitaxial heterostructure for realizing device fabrication. A catalyst-free growth of graphene on an insulating substrate can exclude the problematic transfer process. Hwang et al. presented the van der Waals epitaxial growth of graphene on c-plane (0001) sapphire by CVD without a metal catalyst. Electrical analysis revealed that the Hall mobility at room temperature was above 2000 $cm^2V^{-1}s^{-1}$ [18].

Graphene on metal substrates: The need for metal contacts for the fabrication of graphene-based devices attracted much attention on metal–graphene interfaces. The epitaxial growth of graphene on Co, Ni, Ir, Ru, Cu, Pd, Pt, Rh, Ir, and Ni has been widely reported in the literature. Sutter et al. reported that growth by carbon segregation yielded macroscopic monolayer graphene domains, which extended continuously across Pt (111) substrate steps and were bounded by strongly faceted edges. From the maps of band structure, it was confirmed that the weak substrate coupling was attributed to growth and structural properties. The epitaxial growth of graphene on transition metal surfaces such as Pt (111), Ir (111), Rh (111), and Ni (111) by two different deposition techniques has also been compared and well understood [19].

1.3.3 Chemical Vapor Deposition

Chemical vapor deposition is one of the most promising methods for preparing large-scale, high-quality graphene films [20–23]. Many transition metals are used as catalyst for producing graphene films, such as Ir, Ni, Ru, Cu, and Fe. Among these transition metals, Ni and Cu are widely used, since high-quality, large-area, few-layers, or even single-layer graphene films can be obtained on these substrates.

Kim et al. developed a simple method to grow large-area graphene on pattern Ni substrate and transferred to unspecific substrates, as shown in Fig. 1.6 [21]. Reina et al. produced large-area, high-quality graphene films on poly Ni substrates and discussed the influence of growth parameters, such as growth temperature, growth pressure, growth time, flow rate of carrier gas, heating and cooling speed of the furnace [21].

For copper substrates, almost only single-layer graphene can be obtained. Since copper cannot dissolve large amounts of carbon in it, the growth is a self-limited process. Li et al. first reported large-area, uniform, single-layer graphene on copper foils by CVD [22]. Since then, many reports have emerged. The breakthrough, however, came in 2010 when Bae et al. demonstrated the roll-to-roll production of monolayer 30-inch graphene films, which is cost and time efficient [23]. This method opens a new route to large-scale production of high-quality graphene films for practical applications.

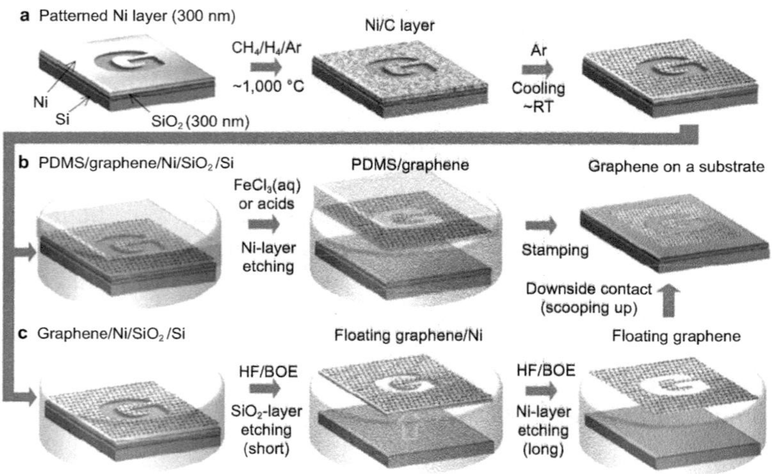

Figure 1.6 Synthesis, etching, and transfer processes for the large-scale and patterned graphene films. (a) Synthesis of patterned graphene films on thin nickel layers. (b) Etching using $FeCl_3$ (or acids) and transfer of graphene films using a PDMS stamp. (c) Etching using buffered oxide etch (BOE) or hydrogen fluoride (HF) solution and transfer of graphene films. Reprinted by permission from Macmillan Publishers Ltd: *Nature*, Ref. 21, Copyright 2009.

Figure 1.7 presents photographs of the roll-based synthesis and transfer process [23]. An 8-inch-wide tubular quartz reactor is used in the 39-inch CVD system, allowing a monolayer graphene film to be synthesized on a roll of copper foil with dimensions as large as 30 inches in the diagonal direction. First, the roll of copper foil is inserted into a tubular quartz tube and then heated to 1000°C with flowing 8 s.c.c.m. H_2 at 90 mtorr (Fig. 1.7a). When the temperature reached 1000°C, the sample is annealed for 30 min without changing the flow rate or pressure; then the gas mixture of CH_4 and H_2 is flowed at 460 mtorr with rates of 24 and 8 s.c.c.m. for 30 min, respectively. Finally, the sample is rapidly cooled to room temperature ($\sim$10°Cs^{-1}) with flowing H_2 under a pressure of 90 mtorr. After being grown on the copper foil, the graphene film is attached to a thermal release tape (Jinsung Chemical Co. and Nitto Denko Co.) by applying light pressure (0.2 MPa) between the two rollers. After etching the copper foil in a plastic bath filled with

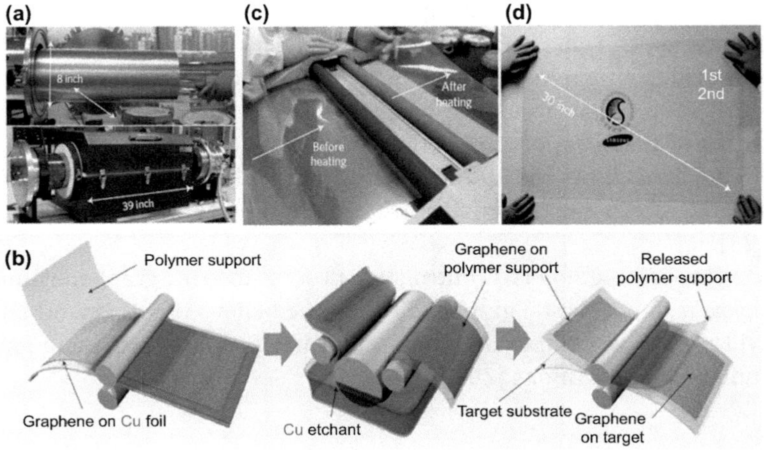

Figure 1.7 Roll-based production of graphene films grown on a copper foil: (a) Copper foil wrapping around a 7.5-inch quartz tube to be inserted into an 8-inch quartz reactor. The lower image shows the stage in which the copper foil reacts with CH_4 and H_2 gases at high temperatures. (b) The process includes adhesion of polymer supports, copper etching (rinsing), and dry transfer-printing on a target substrate. A wet-chemical doping can be carried out using a set-up similar to that used for etching. (c) Roll-to-roll transfer of graphene films from a thermal release tape to a PET film at 120°C. (d) A transparent, ultra-large-area graphene film transferred on a 35-inch PET sheet. Reprinted by permission from Macmillan Publishers Ltd: *Nature Nanotechnology*, Ref. 23, Copyright 2010.

copper etchant, the transferred graphene film on the tape is rinsed with deionized water to remove residual etchant and is then ready to be transferred to any kind of flat or curved surface on demand (Fig. 1.7b). There are three essential steps in the roll-to-roll transfer: (i) adhesion of polymer supports to the graphene on the copper foil; (ii) etching of the copper layers; and (iii) release of the graphene layers and transfer onto a target substrate. In the adhesion step, the graphene film, grown on a copper foil, is attached to a thin polymer film coated with an adhesive layer by passing between the two rollers. In the subsequent step, the copper layers are removed by electrochemical reaction with aqueous 0.1 M ammonium persulfate solution $(NH_4)_2S_2O_8$. Finally, the graphene films are transferred from the polymer support onto a target substrate by removing the adhesive force holding the graphene films. When using thermal release tapes, the graphene films are detached from the tapes and released to counter substrates by thermal treatment (Fig. 1.7c). The third step is not necessary when the target substrate is directly attached to the copper foil in the first step by permanent adhesion. Figure 1.7d shows the 30-inch multilayer graphene film transferred to a roll of 188 mm thick polyethylene terephthalate (PET) substrate [23].

The CVD method is a controllable process to grow graphene film or even single-crystal graphene flakes. We believe that this method will play a very important role in the production of graphene for industrial applications [20–23].

1.3.4 Solution Processing

To obtain a large amount of graphene, solution processing is the method of choice as it allows scaling up easily. The quality of the obtained graphene flakes is very high in accordance with the micromechanical exfoliation. The size of the flakes, however, is very small; neither is the controllability given. The easiest method would be to disperse the graphite in an organic solvent with nearly the same surface energy as graphite [24–35].

1.3.4.1 Graphite/graphene oxides

The first observation of graphite was exfoliated by Schafhaeutl, when H_2SO_4 was used for the intercalation [24]. Since then, a number of chemical species have been found to form intercalated compounds with graphite. Comprehensive reviews on the preparation of dispersions of graphene oxide (GO) platelets and reduced graphene oxide (rGO) platelets, made from GO, have recently appeared. In general, GO is produced by the oxidative treatment of graphite via one of the three principal methods developed by Brodie [25], Hummers [26], and Staudenmeier [27], or some variation of these methods.

Graphite powders may be dispersed in organic solvents such as N-methyl-pyrrolidone (NMP) and sodium dodecylbenzene sulfonate (SDBS) in non-aqueous and aqueous solutions, respectively. Reintercalation of thermally exfoliated expandable graphite with oleum and tetrabutyl-ammonium hydroxide (TBA) has also produced single-layer graphene sheets with a yield of 90% after purification. Even without sonication, ternary potassium salt-intercalated graphite readily gives isolated graphene sheets in NMP [28, 29].

Another typical method involves treating graphite powder with inorganic solvents to obtain graphite oxide. Brodie and Staudenmaier used a combination of potassium chlorate ($KClO_3$) with nitric acid (HNO_3) to oxidize graphite, while the Hummers method involves treatment of graphite with potassium permanganate ($KMnO_4$) and sulfuric acid (H_2SO_4) [30]. The polar oxygen functional groups of GO render it hydrophilic; GO can be dispersed in many solvents, particularly well in water.

1.3.4.2 Reduced graphene oxides

GO surfaces are highly oxidized and contain residual epoxides, hydroxides, and carboxylic acid groups. GO was obtained by the modified Hummers method. GO (100 mg) and ammonium bicarbonate (1 g) were milled to a fine powder. Then this powder was placed into a microwave oven (WP800P23-K, 2450 MHz, 800 W) for 60 s. Under MWI, the precursors responded quickly, accompanied by lightning. After completion of the reaction, the mixture was cooled to room temperature and then washed many times with water until the pH of the filtrate was neutral. The residual black solid was dried under vacuum. The whole method is described in Fig. 1.8 [31]. Various reduction methods have been reported to obtain rGO sheets, such as thermal reduction, chemical reduction, electrochemical reduction, photochemical reduction, and photothermal reduction. Among these methods, thermal, chemical, and electrochemical reductions are more frequently used [31].

Thermal reduction: Reduction via thermal treatment is an efficient and low-cost method [32]. Thermally reduced graphene is obtained by subjecting graphite oxide to a temperature of $\sim$1000°C in Ar or N_2 atmosphere. The resulting product is highly reduced and has the highest C/O atomic ratio among all types of

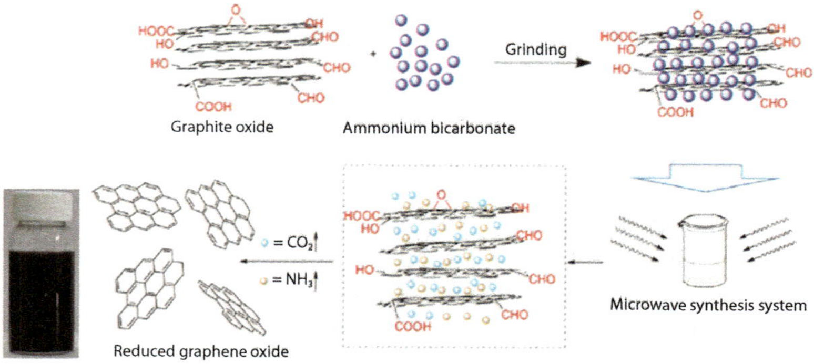

Figure 1.8 Procedures used in the preparation of reduced graphene oxide. Reprinted from Ref. 31, Copyright 2012, with permission from Elsevier.

rGO; the oxidation method of graphite oxide precursor strongly influences the electrochemical properties of the corresponding reduced graphene. However, thermally reduced graphene is also heavily damaged during the exfoliation process, with its structural and electrochemical properties in close resemblance to amorphous carbon [32].

Chemical reduction: To date, chemically reduced graphene constitutes a very wide group of materials [33]. While a variety of chemical reducing agents are used (i.e., N_2H_4, $NaBH_4$, $LiAlH_4$), only a few of them have known mechanisms and target specific oxygen-containing groups. The rGO sheets reduced by hydriodic acid and acetic acid have shown the best electronic conductivity. Due to the less toxicity of hydriodic acid, this method is expected to replace the hydrazine reduction for the mass production of rGO dispersions and fabrication of rGO thin films. Despite the fact that the conductivity of rGO is orders lower compared to the pristine graphene due to the presence of residual oxygenated groups and defects, the reactive surfaces of GO and rGO provide the tunability in electronic and optoelectronic properties via chemical reactions, and the feasibility for composite incorporation [33].

Electrochemical reduction: Electrochemical reduction has been presented as an effective way to remove oxygen functional groups from GO [34, 35]. Electrochemically reduced graphene obtained from GO prepared via the Staudenmaier and Hofmann oxidation methods (using chlorate oxidant) undergoes chemically irreversible reduction, which means that these materials have stable and wide electrochemical windows once they are electrochemically reduced. However, graphene oxides prepared by the Hummers and Tour methods (using permanganate oxidant) lead to reversible chemistry of oxygen-containing groups [34, 35].

1.4 Optical Properties of Graphene

1.4.1 Linear Optical Properties

Due to the unique electronic structure in which conical-shaped conduction and valence bands meet at the Dirac point, the optical conductance of pristine monolayer graphene is frequency independent in a broad range of photon energies:

$$G_1(\omega) = G_0 \equiv e^2/4\hbar \tag{1.1}$$

where ω is the radian frequency, e is electron charge, and p is reduced Planck's constant. As a direct consequence of this universal optical conductance, the optical transmittance of pristine graphene is also frequency independent and solely determined by the fine-structure constant $R = e^2/pc$ (c is the speed of light):

$$T \equiv (1 + 2\pi G/c)^{-2} \approx 1 - \pi\alpha \approx 0.977 \tag{1.2}$$

When scaled to its atomic thickness, graphene actually shows strong broadband absorption per unit mass of the material ($\pi R = 2.3\%$), which is ~50 times higher than GaAs of the same thickness [36, 37]. The reflectance under normal light incidence is relatively weak and written as $R = 0.25\pi^2 R^2 T = 1.3 \times 10^{-4}$, which is much smaller than the transmittance [38]. The absorption of few-layer graphene can be roughly estimated by scaling the number of layers ($T = 1 - N\pi\alpha$). In principle, a low sheet resistance can be attained without sacrificing the properties of transparency too much (~10 Ω Sq^{-1} for T >90%). As such, graphene enthusiasts are hopeful that few-layer graphene can potentially replace indium tin oxide (ITO) as transparent conductors for applications in solar cells and touchscreen in cases when it is sufficiently doped [36, 37].

The property of graphene is largely controlled by the chemical potential (μ, also Fermi level E_F). The transition from dielectric to metallic behavior in terms of optical transition can be tuned by chemical doping or electrical gating [39]. Within the random-phase approximation, the dynamic optical response of graphene can be derived from Kubo formula in a complex form consisting of interband and intraband contributions [40, 41]:

$$\sigma = \sigma_{\text{intra}} + \sigma_{\text{inter}}{}' + i\sigma_{\text{inter}}{}'' \tag{1.3}$$

where the intraband conductivity σ_{intra} has the Drude-like form

$$\sigma_{\text{intra}} = \sigma_0 \frac{4\mu}{\pi} \frac{1}{\hbar\tau_1 - i\hbar\omega} \tag{1.4}$$

where $\sigma_0 = \pi e^2/(2h)$, τ_1 is the relaxation rate associated with intraband transitions, and $\mu > 0$ is the chemical potential. The interband contribution has the form

$$\sigma_{\text{intra}}{}' = \sigma_0 \left(1 + \frac{1}{\pi} \arctan \frac{\hbar\omega - 2\mu}{\hbar\tau_2} - \frac{1}{\pi} \arctan \frac{\hbar\omega + 2\mu}{\hbar\tau_2}\right) \tag{1.5}$$

and

$$\sigma_{\text{intra}}{}'' = -\sigma_0 \frac{1}{2\pi} \ln \frac{(2\mu + \hbar\omega)^2 + \hbar^2\tau_2^2}{(2\mu - \hbar\omega)^2 + \hbar^2\tau_2^2} \tag{1.6}$$

where τ_2 is the relaxation rate associated with interband transitions.

According to the above equations, both interband and intraband conductivities are closely correlated with the chemical potential of graphene and the frequency of incident light. The chemical potential of doped graphene is determined by the carrier concentration $n_0 = (\mu/p\nu)^2/\pi$, which can be controlled via chemical doping or electrical gating. There is no intraband contribution for pristine graphene, which has $\mu = 0$. Due to the occurrence of zero bandgap, the intraband and interband contributions compete and the interband transition only occurs above the threshold of $|\mu| = \hbar\omega/2$, as shown in Fig 1.9. Generally, for slightly doped graphene ($|\mu| < \hbar\omega/2$), the high-frequency dynamic conductivity is dominated by interband contribution for Dirac fermions extending across a wide bandwidth from visible to infrared (IR), which covers the telecommunication bands and also the mid-IR regions. The intraband contribution will only be relevant in the terahertz range when $|\mu| > \hbar\omega/2$; this condition allows for plasmon momentum

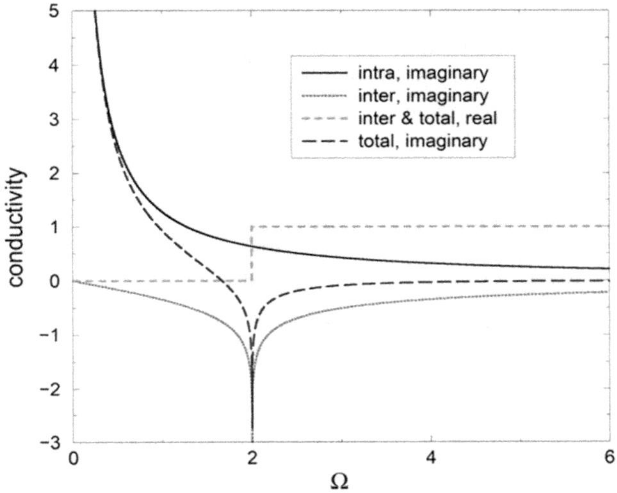

Figure 1.9 The optical conductor rate changes with frequency $\Omega = \omega\hbar/\mu$ (under zero temperature condition and ignoring the effect of collision). Reprinted with permission from Ref. 42, Copyright 2007 by the American Physical Society.

enhancement, and thus propagation of surface plasmon in graphene is possible [42, 43]. As a result, the relative contributions from optical interband transition and intraband transition can be varied, leading to tunable optical response as well as polarization-selective coupling.

1.4.2 Nonlinear Optical Properties

The optical nonlinearities originate from the interactions of the optical field with electrons and phonons. Specifically, the electric field of the optical signal resonates with the electrons in the outer shells of carbon atoms and displaces the electron cloud with respect to the nucleus, resulting in a net dipole moment (polarization). In the simplest case of linear optics regime, the induced displacement (χ) manifests as electronic polarization (P), obeying a linear relationship with the applied field (E):

$P = \varepsilon_0 \chi^{(1)} E$, where ε_0 is the permittivity of free space and $\chi^{(1)}$ is the first-order (linear) susceptibility. In the case of a large displacement caused by a strong optical field, electronic polarization is significantly nonlinear with respect to the displacement as well as the applied electric field, which could be described by a Taylor expansion [44]:

$$P = \varepsilon_0 \chi^{(1)} E + \varepsilon_0 \chi^{(2)} E^2 + \varepsilon_0 \chi^{(3)} E^3 + \cdots + \varepsilon_0 \chi^{(n)} E^n + \cdots$$
$$(n = 1, 2, 3, \ldots n, \ldots) \tag{1.7}$$

where $\chi^{(2)}$ and $\chi^{(3)}$ are the second- and third-order nonlinear susceptibilities, associated with nonlinear phenomena such as two-photon absorption (TPA), self-focusing, self-phase modulation, saturable absorption (SA), and optical bistability.

Dipole excitations with bound and free electrons induced by a single photon are described by the first-order susceptibility $\chi^{(1)}$, where the real part is correlated to the real part of the refractive index and the imaginary part is related to optical loss or gain. The effective linear susceptibility to optical fields at a frequency ω can be modulated by applying a dc field across graphene, which changes the refractive index. This linear electro-optic effect is manifested readily in graphene, which forms the basis of a broadband optical modulator recently demonstrated [45]. Due to the inversion symmetry of the graphene unit cell, the second-order susceptibility $\chi^{(2)}$ is generally absent as a result of the cancellation of the optically excited second-order electrical current at opposite momenta with respect to the Dirac point. However, $\chi^{(2)}$ may be operative in strained [46], disordered, or functionalized graphene in which the bilayer symmetry is broken, thus allowing many possibilities to be exploited in $\chi^{(2)}$ nonlinear electro-optic active material [36, 37]. For example, in graphene derivatives that do not possess inversion symmetry, the application of an optical field at frequency ω may generate its second harmonic at 2ω, which may find important uses for laser frequency doubling as well as high-resolution optical microscopy. More interesting effects may arise by applying the incident field with two frequencies (ω_1 and ω_2), for example, to generate the sum and difference frequencies ($\omega_{1\pm}\omega_2$).

Recently, Wu et al. [47] demonstrated that a giant and tunable second-order nonlinear susceptibility can be observed in bilayer graphene when an in-plane electrical current between source and drain electrodes is introduced to suppress the cancellation of optically induced electric current. In principle, third-order nonlinear processes are allowed in all materials, whether they process inversion symmetry or not. Most of the optical nonlinearities in graphene are associated with the third-order susceptibility $\chi^{(3)}$. However, $\chi^{(3)}$ is correlated to the polarization per unit volume to the third power of the electric field E and graphene is an infinitesimally thin and isotropic conductivity surface. The classical model fails to fully interpret the optical nonlinearities in graphene, and it is more appropriate to describe the nonlinear response in terms of the nth order total integrated sheet current [48, 49]:

$$J_n^v = \frac{1}{4\pi^2} \int dP_{J_n}^v N(\varepsilon) \tag{1.8}$$

where the thermal factor

$$N(\varepsilon) = n_F(-\varepsilon) - n_F(\varepsilon) = \tanh(\varepsilon/2k_B T)$$

and $J_n^v = \psi \widehat{V}_v \psi$, where the current operator

$$\widehat{V}_v = (\partial H)/(\partial P_v),\ V = x, y$$

The first-order total current, that is, linear optical response, $J_1 = e^2 E/(4\hbar)$, after converting into real units, coincides with the universal conductance deduced from the above-mentioned Kubo formula (Eq. 1.2). For a symmetric crystal,

$$V(x) = V(-x),$$

which implies the second-order total sheet current $J_2 = 0$. The third-order current can be written as [50]:

$$J_3 = J_3(\omega) + J_3(3\omega) = \frac{\sigma_1 e^2 v_F^2 E_0^2}{\hbar^2 \omega^4}[N_1(\omega)e^{j\omega t} + N_3(\omega)e^{3j\omega t}] \tag{1.9}$$

where $v_{F\approx c/300}$, the thermal factors $N_1(\omega) = N(\omega)$, and $\sigma_1 = e^2/4\hbar$

$$N_3(\omega) = 13N(\omega/2)/48 - N(\omega)/3 + 45N(3\omega/2)/48$$

According to Eq. 1.8, the third-order current is a superposition of two third-order terms correlated to three-photon processes, that is, a single frequency term $J_3(\omega)$ and a third-harmonic generation term $J_3(3\omega)$. Both of these two terms are inversely proportional to ω^4 and proportional to E_0, which is the power of the field. They are responsible for a wide variety of optical nonlinear effects in graphene, which includes saturable absorption [51], self-focusing and nonlinear refractive index change (i.e., Kerr effect) [52], optical bistability and switching [53], and soliton propagation [54, 55].

References

1. Geim, A. K., and Novoselov, K. S. (2007). The rise of graphene, *Nat. Mater.*, **6**, pp. 183–191.
2. Geim, A. K., and Kim, P. (2008). Carbon wonderland, *Sci. Am.*, **298**, pp. 90–97.
3. Kroto, H. W., Heath, J. R., O'Brien, S. C., Curl, R. F., and Smalley, R. E. (1985). C60: Buckminsterfullerene, *Nature*, **318**, pp. 162–163.
4. Iijima, S. (1991). Helical microtubules of graphitic carbon, *Nature*, **354**, pp. 56–58.
5. Novoselov, K. S., Geim, A. K., Morozov, S. V., Jiang, D., Zhang, Y., Dubonos, S. V., Grigorieva, I. V., and Firsov, A. A. (2004). Electric field in atomically thin carbon films, *Science*, **306**, pp. 666–669.
6. Morozov, S. V., Novoselov, K. S., Katsnelson, M. I., Schedin, F., Elias, D. C., Jaszczak, J. A., and Geim, A. K. (2008). Giant intrinsic carrier mobilities in graphene and its bilayer, *Phys. Rev. Lett.*, **100**, 016602.
7. Du, X., Skachko, I., Barker, A., and Andrei, E. Y. (2008). Approaching ballistic transport in suspended graphene, *Nat. Nanotechnol.*, **3**, pp. 491–495.
8. Lee, C., Wei, X., Kysar, J. W., and Hone, J. (2008). Measurement of the elastic properties and intrinsic strength of monolayer graphene, *Science*, **321**, pp. 385–388.
9. Balandin, A. A., Ghosh, S., Bao, W., Calizo, I., Teweldebrhan, D., Miao, F., and Lau, C. N. (2008). Superior thermal conductivity of single-layer graphene, *Nano Lett.*, **8**, pp. 902–907.
10. Bonaccorso, F., Sun, Z., Hasan, T., and Ferrari, A. (2010). Graphene photonics and optoelectronics, *Nat. Photonics*, **4**, pp. 611–622.
11. Özçelik, V. O., Cahangirov, S., and Ciraci, S. (2012). Epitaxial growth mechanisms of graphene and effects of substrates, *Phys. Rev. B*, **85**, pp. 235456.
12. Huang, H., Chen, W., Chen, S., and Wee, A. T. S. (2008). Bottom-up growth of epitaxial graphene on 6H-SiC(0001), *ACS Nano*, **2**, pp. 2513–2518.
13. Gaskill, D. K., Jernigan, G., Campbell, P., Tedesco, J. L., Culbertson, J., VanMil, B., Myers-Ward, R. L., Eddy, C., Moon, J., and Curtis, D. (2009). Epitaxial graphene growth on SiC wafers, *ECS Trans.*, **19**, pp. 117–124.
14. First, P. N., de Heer, W. A., Seyller, T., Berger, C., Stroscio, J. A., and Moon, J.-S. (2010). Epitaxial graphenes on silicon carbide, *MRS Bulletin*, **35**, pp. 296.

15. Yang, W., Chen, G., Shi, Z., Liu, C.-C., Zhang, L., Xie, G., Cheng, M., Wang, D., Yang, R., and Shi, D. (2013). Epitaxial growth of single-domain graphene on hexagonal boron nitride, *Nat. Mat.*, **12**, pp. 792–797.

16. Dean, C., Young, A., Meric, I., Lee, C., Wang, L., Sorgenfrei, S., Watanabe, K., Taniguchi, T., Kim, P., and Shepard, K. (2010). Boron nitride substrates for high-quality graphene electronics, *Nat. Nanotechnol.*, **5**, pp. 722–726.

17. Xue, J., Sanchez-Yamagishi, J., Bulmash, D., Jacquod, P., Deshpande, A., Watanabe, K., Taniguchi, T., Jarillo-Herrero, P., and LeRoy, B. J. (2011). Scanning tunneling microscopy and spectroscopy of ultra-flat graphene on hexagonal boron nitride, *Nat. Mat.*, **10**, pp. 282–285.

18. Hwang, J., Kim, M., Campbell, D., Alsalman, H. A., Kwak, J. Y., Shivaraman, S., Woll, A. R., Singh, A. K., Hennig, R. G., and Gorantla, S. (2012). Van der Waals epitaxial growth of graphene on sapphire by chemical vapor deposition without a metal catalyst, *ACS Nano*, **7**, pp. 385–395.

19. Chen, H., Zhu, W., and Zhang, Z. (2010). Contrasting behavior of carbon nucleation in the initial stages of graphene epitaxial growth on stepped metal surfaces, *Phys. Rev. Lett.*, **104**, pp. 186101.

20. Reina, A., Jia, X., Ho, J., Nezich, D., Son, H., Bulovic, V., Dresselhaus, M. S., and Kong, J. (2008). Large area, few-layer graphene films on arbitrary substrates by chemical vapor deposition, *Nano Lett.*, **9**, pp. 30–35.

21. Kim, K. S., Zhao, Y., Jang, H., Lee, S. Y., Kim, J. M., Kim, K. S., Ahn, J.-H., Kim, P., Choi, J.-Y., and Hong, B. H. (2009). Large-scale pattern growth of graphene films for stretchable transparent electrodes, *Nature*, **457**, pp. 706–710.

22. Li, X., Cai, W., An, J., Kim, S., Nah, J., Yang, D., Piner, R., Velamakanni, A., Jung, I., and Tutuc, E. (2009). Large-area synthesis of high-quality and uniform graphene films on copper foils, *Science*, **324**, pp. 1312–314.

23. Bae, S., Kim, H., Lee, Y., Xu, X., Park, J.-S., Zheng, Y., Balakrishnan, J., Lei, T., Kim, H. R., and Song, Y. I. (2010). Roll-to-roll production of 30-inch graphene films for transparent electrodes, *Nat. Nanotechnol.*, **5**, pp. 574–578.

24. Schafhaeutl, C. (1840). Ueber die Verbindungen des Kohlenstoffes mit Silicium, Eisen und anderen Metallen, welche die verschiedenen Gallungen von Roheisen, Stahl und Schmiedeeisen bilden, *Journal für Praktische Chemie*, **21**(1), pp. 129–157.

25. Kobayashi, Y., Fukui, K.-I., Enoki, T., Kusakabe, K., and Kaburagi, Y. (2005). Observation of zigzag and armchair edges of graphite using scanning tunneling microscopy and spectroscopy, *Phys. Rev. B*, **71**(19), pp. 193406.

26. Nakada, K., Fujita, M., Dresselhaus, G., and Dresselhaus, M. S. (1996). Edge state in graphene ribbons: Nanometer size effect and edge shape dependence, *Phys. Rev. B*, **54**(24), pp. 17954.
27. Son, Y.-W., Cohen, M. L., and Louie, S. G. (2006). Half-metallic graphene nanoribbons, *Nature*, **444**(7117), pp. 347–349.
28. Hernandez, Y., Nicolosi, V., Lotya, M., Blighe, F. M., Sun, Z., De, S., McGovern, I., Holland, B., Byrne, M., and Gun'Ko, Y. K. (2008). High-yield production of graphene by liquid-phase exfoliation of graphite, *Nat. Nanotechnol.*, **3**(9), pp. 563–568.
29. Vallés, C., Drummond, C., Saadaoui, H., Furtado, C. A., He, M., Roubeau, O., Ortolani, L., Monthioux, M., and Pénicaud, A. (2008). Solutions of negatively charged graphene sheets and ribbons, *J. Am. Chem. Soc.*, **130**(47), pp. 15802–15804.
30. Ehrburger, P., and Donnet, J. (1973). Etude de l'oxydation des carbones en oxyde graphitique, *Carbon*, **11**(4), pp. 309–316.
31. Shen, J., Li, T., Lang, Y., Shi, M., Li, N., and Ye, M. (2012). One-step solid state prepration of reduced graphene oxide, *Carbon*, **50**, pp. 2134–2140.
32. Mao, S., Lu, G., Yu, K., Bo, Z., and Chen, J. (2010). Specific protein detection using thermally reduced graphene oxide sheet decorated with gold nanoparticle conjugates, *Adv. Mat.*, **22**(32), pp. 3521–3526.
33. Li, D., Mueller, M. B., Gilje, S., Kaner, R. B., and Wallace, G. G. (2008). Processable aqueous dispersions of graphene nanosheets, *Nat. Nanotechnol.*, **3**(2), pp. 101–105.
34. Guo, H.-L., Wang, X.-F., Qian, Q.-Y., Wang, F.-B., and Xia, X.-H. (2009). A green approach to the synthesis of graphene nanosheets, *ACS Nano*, **3**(9), pp. 2653–2659.
35. Wang, Z., Zhou, X., Zhang, J., Boey, F., and Zhang, H. (2009). Direct electrochemical reduction of single-layer graphene oxide and subsequent functionalization with glucose oxidase, *J. Phys. Chem. C*, **113**(32), pp. 14071–14075.
36. Kuzmenko, A., Van Heumen, E., Carbone, F., and van der Marel, D. (2008). Universal optical conductance of graphite, *Phys. Rev. Lett.*, **100**, pp. 117401.
37. Stauber, T., Peres, N., and Geim, A. (2008). Optical conductivity of graphene in the visible region of the spectrum, *Phys. Rev. B*, **78**, pp. 085432.
38. Nair, R., Blake, P., Grigorenko, A., Novoselov, K., Booth, T., Stauber, T., Peres, N., and Geim, A. (2008). Fine structure constant defines visual transparency of graphene, *Science*, **320**, pp. 1308–1308.

39. Li, Z., Henriksen, E. A., Jiang, Z., Hao, Z., Martin, M. C., Kim, P., Stormer, H., and Basov, D. N. (2008). Dirac charge dynamics in graphene by infrared spectroscopy, *Nat. Phys.*, **4**, pp. 532–535.
40. Gusynin, V., and Sharapov, S. (2006). Transport of Dirac quasiparticles in graphene: Hall and optical conductivities, *Phys. Rev. B*, **73**, pp. 245411.
41. Gusynin, V., Sharapov, S., and Carbotte, J. (2006). Unusual microwave response of Dirac quasiparticles in graphene, *Phys. Rev. Lett.*, **96**, pp. 256802.
42. Mikhailov, S., and Ziegler, K. (2007). New electromagnetic mode in graphene, *Phys. Rev. Lett.*, **99**, 016803.
43. Koppens, F. H., Chang, D. E., and Garcia de Abajo, F. J. (2011). Graphene plasmonics: A platform for strong light–matter interactions, *Nano Lett.*, **11**, pp. 3370–3377.
44. Banerjee, P. P. (2003). *Nonlinear Optics: Theory, Numerical Modeling, and Applications*, CRC Press.
45. Liu, M., Yin, X., Ulin-Avila, E., Geng, B., Zentgraf, T., Ju, L., Wang, F., and Zhang, X. (2011). A graphene-based broadband optical modulator, *Nature*, **474**, pp. 64–67.
46. Mishchenko, E. (2009). Dynamic conductivity in graphene beyond linear response, *Phys. Rev. Lett.*, **103**, pp. 246802.
47. Wu, S., Mao, L., Jones, A. M., Yao, W., Zhang, C., and Xu, X. (2012). Quantum-enhanced tunable second-order optical nonlinearity in bilayer graphene, *Nano Lett.*, **12**, pp. 2032–2036.
48. López-Rodríguez, F., and Naumis, G. (2008). Analytic solution for electrons and holes in graphene under electromagnetic waves: Gap appearance and nonlinear effects, *Phys. Rev. B*, **78**, pp. 201406.
49. Ishikawa, K. L. (2010). Nonlinear optical response of graphene in time domain, *Phys. Rev. B*, **82**, pp. 201402.
50. Wright, A., Xu, X., Cao, J., and Zhang, C. (2009). Strong nonlinear optical response of graphene in the terahertz regime, *Appl. Phys. Lett.*, **95**, pp. 072101.
51. Bao, Q., Zhang, H., Wang, Y., Ni, Z., Yan, Y., Shen, Z. X., Loh, K. P., and Tang, D. Y. (2009). Atomic-layer graphene as a saturable absorber for ultrafast pulsed lasers, *Adv. Funct. Mat.*, **19**, pp. 3077–3083.
52. Zhang, H., Virally, S., Bao, Q., Kian Ping, L., Massar, S., Godbout, N., and Kockaert, P. (2012). Z-scan measurement of the nonlinear refractive index of graphene, *Opt. Lett.*, **37**, pp. 1856–1858.

53. Bao, Q., Chen, J., Xiang, Y., Zhang, K., Li, S., Jiang, X., Xu, Q. H., Loh, K. P., and Venkatesan, T. (2015). Graphene nanobubbles: A new optical nonlinear material, *Adv. Opt. Mat.*, **3**(6).

54. Belonenko, M., Lebedev, N., and Tuzalina, O. Y. (2009). Electromagnetic solitons in a system of graphene planes with Anderson impurities, *J. Russian Laser Res.*, **30**, pp. 101–108.

55. Zhang, H., Tang, D., Zhao, L., Bao, Q., and Loh, K. P. (2010). Vector dissipative solitons in graphene mode locked fiber lasers, *Opt. Commun.*, **283**, pp. 3334–3338.

56. Novoselov, K. S., and Castro Neto, A. H. (2012). Two dimensional crystals based heterostructures: materials with tailored properties, *Physica Scripta*, T146.

Chapter 2

Application of Graphene in Lasers

Yao Chen,[a] Haoran Mu,[a] Yupeng Zhang,[b] and Qiaoliang Bao[a,b]

[a]*Institute of Functional Nano & Soft Materials (FUNSOM), Soochow University, 199 Ren-Ai Road, Suzhou Industrial Park, Suzhou Jiangsu 215123, China*
[b]*Department of Materials Science and Engineering, Monash University, 22 Alliance Lane, Building 69, Clayton, 3800, Victoria, Australia*
qiaoliang.bao@gmail.com

2.1 Introduction

Q-switching and mode-locking are mainly two methods to realize ultrashort pulse laser. Q-switching could generate output pulse with width in the ms and ns, which has the characteristics of high-pulse energy. The mode-locking technology can generate ps pulse and fs pulse as well as higher peak power. To achieve Q-switching and mode-locking, both active and passive methods are available. Relatively, passive mode-locking and passive Q-switching methods do not need an electric field or light field modulation; they only need the insertion of nonlinear optical components known as saturable absorbers in the laser cavity, so that it is more convenient and efficient and could be easily realized.

Graphene Photonics, Optoelectronics, and Plasmonics
Edited by Qiaoliang Bao, Hui Ying Hoh, and Yupeng Zhang

ISBN 978-981-4745-60-4 (Hardcover), 978-1-315-19667-1 (eBook)
www.panstanford.com

The performance of a saturable absorber may be measured by the following factors: working wavelength range, modulation depth, loss of unsaturated, saturated light intensity, and the thermal damage threshold value. Because graphene has the characteristics of super wide spectrum scope of work range, femtosecond ultrafast electronic relaxation time (~150 fs), and high thermal damage threshold, it has garnered great interest as a new saturable absorber material. Figure 2.1 is the electronic transition diagram after the absorption of incident light of single-layer graphene. Upon illumination, electrons in the valence band of graphene (orange) are excited to the conduction band (yellow). These hot electrons are cooled in a very short time (10~150 fs), forming the Fermi–Dirac distribution within the band (Fig. 2.1b) with electron temperature T_e. The newly formed electron–hole pairs could block some

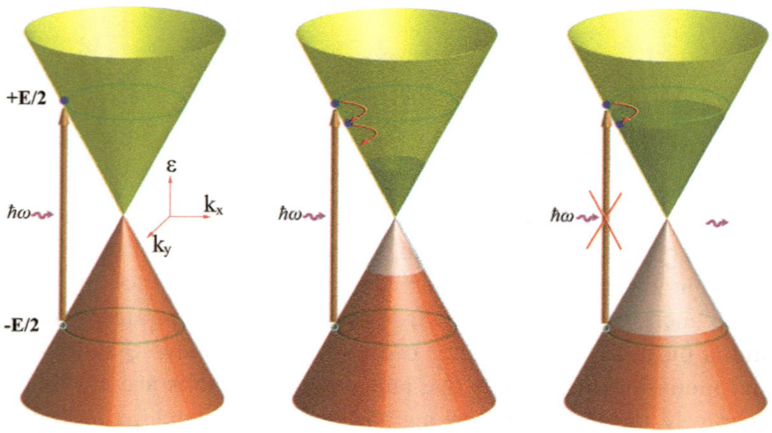

Figure 2.1 Absorption of light in graphene. (a) Schematic excitation process responsible for the absorption of light in graphene. The arrow indicates the optical interband transition. (b) The photogenerated carriers thermalize and cool down within sub-picoseconds to form a hot Fermi–Dirac distribution. An equilibrium electron and hole distribution could be finally approached through the intraband phonon scattering and electron–hole recombination. (c) At high excitation intensity, the photogenerated carriers cause the states near the edge of the conduction and valence bands to fill, blocking further absorption. Reprinted from Ref. 1 with permission from John Wiley and Sons.

interband optical transitions, which were originally possible, within the range of $K_B T_e$ (K_B being the Boltzmann constant) from the Fermi energy E_F. As a result, the absorption of photons is reduced by around $K_B T_e$. In the next 1 ps, with the scattering effect of photon in the band, the hot electron emitted would continue to be cooled. Then the process of compound of electronic hole would dominate. The process describes the linear optical transition process when the incident light intensity is lower. However, with the increase in the incident light intensity, the number of photo-excited electrons and holes significantly increases, and its concentration is higher than the intrinsic graphene concentration at room temperature. These excited electrons and holes would fill the energy state between the graphene conduction band and the valence band, according to the Pauli exclusion principle; the energy level will be unable to accommodate more electrons or holes, thereby blocking the further absorption of light in graphene, manifesting the saturation absorption characteristics (Fig. 2.1c) [1]. In 2009, the first pulse laser based on a graphene saturable absorber was reported. So far, research on graphene saturable absorber lasers has made great progress.

2.2 Mode-Locking Fiber Laser Based on Graphene

Fiber laser has numerous advantages, such as good beam quality, miniaturization and intensification of device structure, easy to heat, and easy to integrate into the optical fiber communication system. Thus, it has become the light source in scientific research and production. Therefore, the improvement of the mode-locked fiber laser has gathered increased attention.

Many ways are available for integrating a graphene saturable absorber into the laser cavity; the most commonly used is the sandwich structure. This method involves transferring the graphene onto the fiber end face of the FC/PC and then connecting with another FC/PC through the connector and integrating into the laser cavity. This method is simple and shortens the length of the

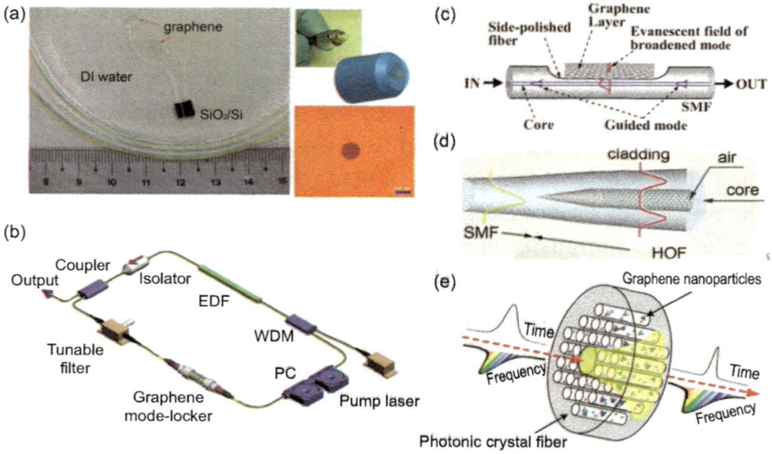

Figure 2.2 Schematic of graphene-based saturable absorber device. (a) Graphene flake floating on the water surface for wet transfer (left); photograph of fiber pigtail coated with graphene film (top right) and optical image of the large-area graphene film covering on the fiber pinhole (bottom right). Reprinted from Ref. 1 with permission from John Wiley and Sons. (b) Circular laser cavity based on graphene mode-locked device with sandwiched structure [5]. (c) Fiber mode-locker with graphene on a side-polished fiber. Reprinted from Ref. 2 with the permission of AIP Publishing. (d) Schematic of a nonlinear interaction of graphene in hollow optical fiber (HOF). Reprinted from Ref. 3 with permission from The Optical Society. (e) Schematic of an evanescent-wave mode-locked pulse propagation through the PCF doped with graphene nanoparticles in hole-cladding region. Reprinted from Ref. 4 with permission from The Optical Society.

cavity caused by the insertion of mode-locking devices. It helps to form a mode-locked pulse with high repetition rate. Figures 2.2a,b show the preparation method of graphene mode-locking devices with a sandwich structure and an example of laser cavity using mode-locking devices. However, in this way, the graphene saturable absorber is placed vertical to the light path; this requires a sample with a high thermal damage threshold. In dealing with a high-power pulse, too much high power would breakdown the graphene easily. To solve these problems, scientists have proposed different methods to couple graphene into the fiber laser cavity, including using D-

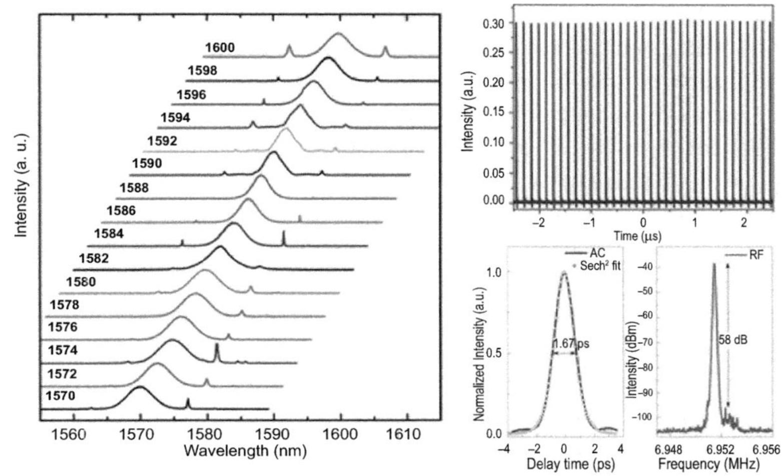

Figure 2.3 Output characteristics of graphene-based wideband tunable mode-locked laser. (a) Wideband output spectra from 1570 nm to 1600 nm; (b) output pulse train; (c) autocorrelation trace and sech 2 fitting curve; (d) radio frequency spectrum of mode-locking pulses. Reprinted from Ref. 7 with permission from John Wiley and Sons.

shaped fiber or tapered fiber coupling side instantaneous field [2], hollow fiber absorber filled with graphene [3], and embedding the graphene nano-particle in the photonic crystal fiber [4], as shown in Figs. 2.2c–e.

At present, a large number of mode-locked fiber lasers based on graphene have been developed. In 2010, Sun, Bao, and Zhang, respectively, prepared fiber mode-locking lasers, which work at 1550 nm with a 30 nm adjustable range in the negative dispersion and positive dispersion cavity [5–7].

Figure 2.3a depicts the spectrum of a 1570~1600 nm waveband continuous adjustable laser, and Figs. 2.3b–d show the respective laser pulse sequence diagrams, autocorrelation spectrum, and RF spectrum. From the figure, it can be seen that for the laser output typical soliton pulse, the pulse width is 1.67 ps and the signal-to-noise ratio is greater than 58 dB. It is worth noting that the 30 nm wideband adjustable range surpasses the traditional index saturable absorption devices achieve (SESAM adjustable range is about 10 nm). Then, the authors made soliton mode-locked optical fiber laser, which can work on positive and negative dispersion and zero dispersion state, and the adjustable range is 30 nm. In terms of output power, the negative dispersion erbium-doped fiber laser cavity usually generates only a few tens of milliwatts pulse and pulse power is lower too. At present, the maximum energy of the single pulse is 7.3 nJ, and the result is obtained by 3.5 W high pump power. As a result, an ytterbium-doped fiber with higher light conversion efficiency is used. The positive dispersion cavity, in which double gold Sulzberger–Landau mechanism replaces traditional nonlinear Schrödinger mechanism, allows higher chirp pulse power, and output power of mode-locked pulse is greatly increased. Currently, the mode-locked pulse energy obtained by this method reached 10.2 nJ at best. With a peripheral circuit, the chirp pulse amplification to the mode-locked pulse is another important method to achieve high-energy output pulse. Sobon et al. [8] used an additional light path to achieve the chirp pulse amplification of graphene mode-locked femtosecond laser output. They obtained 1 W average output power and 20 nJ single pulse energy without influencing the beam quality.

In order to give full play to the graphene ultrafast electronic relaxation and achieve ultrashort mode-locked pulses, in 2011, Cunning et al. [8] made a graphene saturable absorption mirror of low loss by depositing pure graphene polymer on the mirror plated with a 250 nm gold film and using the linear optical fiber cavity design, which shortened the width of the graphene mode-locked laser to below 200 fs. For improving the output pulse repetition rate, the harmonic mode-locking can be dozens of times. Currently, 1 μm and 1.5 μm band harmonic mode-locked graphene fibers have been successfully implemented, and the high repetition rate reached 2.22 GHz.

2.3 Q-Switching Fiber Laser Based on Graphene

By the Q-switching technology, the optical pulse width can reach the level of ns and the peak power can reach the level of MW. Compared with the optical fiber mode-locking technology, using the technology of fiber laser Q-switching can produce short pulse of more energy. Graphene has considerable modulation depth, and the characteristics of working in wideband fully meet the requirements of making Q-switching devices. Being similar to a fiber mode-locked laser, the sandwich structure is the main way to integrate graphene into the Q-switching fiber laser cavity. Usually, graphene is inserted between two FC/PC optical fiber connectors, using the tapered fiber coupling graphene to obtain high-power pulse output.

At present, 1, 1.5, 2, and even 2.78 μm of graphene Q-switching fiber lasers have been reported. At the same time, wavelength tunable graphene Q-switching fiber lasers have been successfully built, with the largest tuning range up to 50 nm. With in-depth research, the performance of graphene Q-switching fiber laser has enhanced unceasingly. Liu et al. deposited oxidized graphene on a tapered optical fiber, and then fixed it in a u-shaped base, placing it in a linear laser cavity as a high-energy saturable absorber, with double-cladding thulium-doped fiber as the gain medium, so as to realize a high-power pulse output [9].

Figure 2.4a shows the laser cavity structure, and Fig. 2.4b shows the typical Q output characteristic curve. The pulse Q-switching center wavelength is 2030 nm; the repetition rate is 45 kHz; and the pulse width is 3.8 mu with the pump output power being 6.27 W. The laser conversion efficiency reached 21.8%, the single pulse output reached 6.71 μJ, and the power reached 302 mW in the 2 μm band, which is the graphene Q-switching fiber laser of the highest output power and the largest single pulse energy. Graphene showed superior performance in the optical Q-switching fiber laser, which fully shows its advantages in the preparation of a Q-switching laser and offers a new way to build a practical application of high-energy laser.

2.4 Graphene in the Solid State Laser

In 2010, Tan et al. [10] conducted graphene mode-locked solid state laser research and achieved a mode-locked pulse of 4 ps width at 1064 nm, by depositing the graphene on the quartz substrate and as a saturable absorber into solid Nd:YAG laser. In 2011, Xu

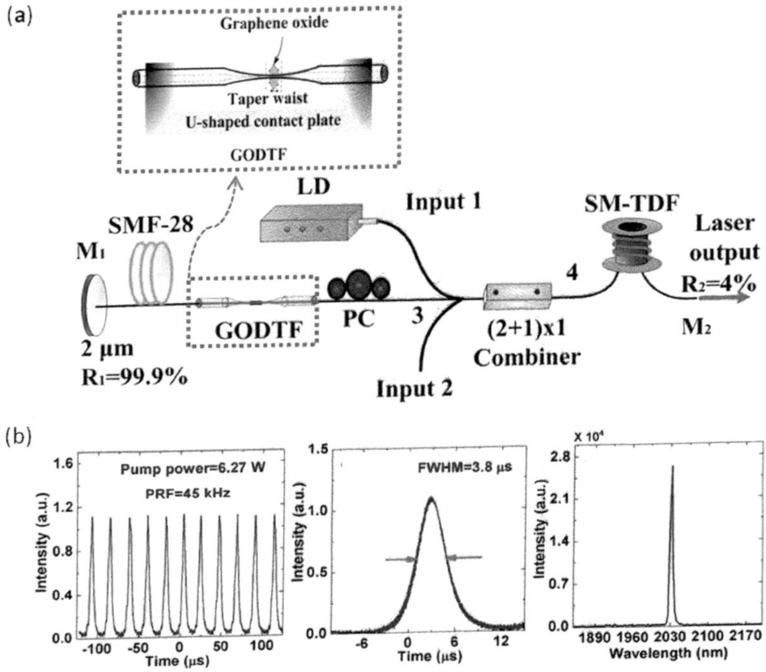

Figure 2.4 Output characteristics of a high-energy graphene-based Q-switched fiber laser. (a) Schematic of the laser cavity and (b) output pulse train, output spectrum, and a single pulse. Reprinted from Ref. 7 with permission from John Wiley and Sons.

et al. [11] adopted the method of spin coating and deposited a multilayer graphene slice of large size (20 microns) on several layers of $SiO_2/TiO_2/BK_7$ glass substrate, forming a saturated absorption mirror, placed into Nd:Gd-VO_4 solid laser and obtained a mode-locked pulse of 16 ps width at 1065 nm. Then, Cho et al. [12] reduced its linear light absorption loss using single-layer graphene, and achieved a high-performance mode-locked pulse of working wavelength 1.25 μm, pulse width less than 100 fs, and average output power being 230 mW. Figure 2.5a gives the structure of solid state laser cavity, which generated a typical mode-locked output spectrum. The fundamental frequency repetition rate is about 88.15 MHz, the signal-to-noise ratio is more than 65 dBm, and the pulse

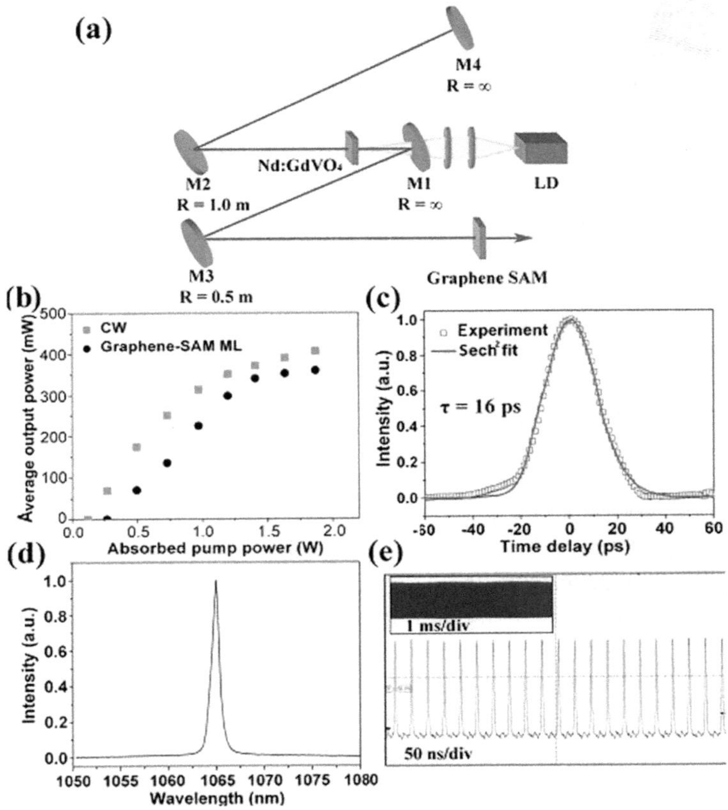

Figure 2.5 Schematic and output characteristics of graphene-based mode-locked solid laser. (a) Experimental setup of the mode-locked laser. LD–laser diode; M1–input mirror; M2 and M3–fold mirrors; M4–end mirror. (b) Average output power versus absorbed pump power in CW and graphene-SAM mode-locking (ML) regimes. (c) Autocorrelation trace of the 16 ps pulse. (d) Emission spectrum. (e) Mode-locked pulse trains recorded in 50 ns and 1 ms per division (div) time scales. Reprinted from Refs. 9 and 11 with permission from The Optical Society.

width is about 91 fs, as shown in Figs. 2.5b–d. In 2012, Liu et al. [13] utilized mode-locking in Tm:$YAlO_3$ laser using oxidized graphene, achieving laser pulse work at 2 μm, the average output power being 260 mW. Baek et al. [14] extended the working wavelength of graphene solid state laser to 800 nm and achieved a width of laser pulse less than 70 fs. In 2013, Cafiso utilized single-layer graphene to get mode-locked in Cr: YAG laser, the working wavelength was 1.5 mm, pulse width was 91 fs, average output power was more than 100 mW [15].

At the same time, research on graphene Q-switching solid laser has made certain progress. Yu et al. [16] achieved Q-switching in Nd:YAG laser using graphene grown with the SiC epitaxial method, achieving output pulse of 159.2 nJ and pulse width of 161 ns. Li et al. achieved a high-energy output pulse, with a single pulse energy of 3.2 μJ [17]. Xu et al. [11] prepared the modulation depth of graphene close to 100%, which used the chemical vapor deposition method. The average output power is up to 1.6 W. The research results undoubtedly show a great potential for graphene in the ultrafast optical field.

At present, graphene has been applied to solid state lasers such as Ti:Sapphire, Nd:KLu $(WO_4)_2$ [11], Nd:YAG, Nd:$GdVO_4$, Nd:$GdVO_4$, Cr:YAG, Cr^{4+}:forsterite, Tm, LSO [15], and Cr:ZnS. The output pulse covers the wavelength of 800 nm, 1 μm, 1.25 μm, 1.4 μm, 1.5 μm, and 2 μm [15]. The maximum output power reached to the level of watt, and the output pulse width was less than 100 fs. These parameters showed the huge success of the application of graphene in solid state laser fields.

Both optical fiber laser and solid laser have their advantages and disadvantages. Fiber laser has the characteristics of small volume and good beam quality, and fiber design is easy to be integrated into the system, suitable for light communication, scientific research, industrial welding, and medical and other fields. Solid state laser has higher output power and output energy as well as narrower pulse width. It is widely applied in the fields of military, processing, medical, and scientific research. Optical fiber laser uses fiber doped with rare earth elements as the gain medium, whereas solid state laser usually uses doping ions or other active substances of transparent crystal as the gain medium. The fluorescence quantum efficiency of this kind of gain medium is higher and is more likely to produce a high-power laser output and better beam quality. Because of the large size, it is not easy to be integrated into the fiber system and is susceptible to interference by the outside environment.

Therefore, solid state lasers are more suitable for the environment, which are stable and require higher beam quality.

2.5 Conclusion

With the improvement in production process and properties, the indicators of lasers based on graphene also constantly improved, embodied as extended working wavelength, bigger output power, shorter pulse width, and higher repetition rate. Given the progress achieved in utilizing graphene in the field of ultrafast laser in just a few years, it can be seen that graphene is of great potential and has good prospects in the field. But the application of graphene in lasers exposed itself to some inherent defects.

First, although the linear absorption spectra of graphene are smooth and continuous, this does not mean that the graphene saturation threshold is the same in each band. Experimental results show that the saturation threshold of graphene in the long wavelength region (such as far infrared) is lower than its saturation threshold in the near-infrared and visible light, which means that graphene may be more suitable for a band above 1.5 μm.

Second, the modulation depth of the monolayer graphene is very low. The modulation depth can be improved if the number of graphene layers is increased. Unsaturated optical loss is increased with an increase in the number of layers. These problems can be solved in two ways: One way is integrating graphene and metal to build a graphene–metal saturable absorption mirror similar to SESAM, with external grid voltage, adjusting the modulation depth of graphene, combining active mode-locking and passive mode-locking to improve the saturable absorption performance of graphene. In addition, other two-dimensional materials have received wide attention with the rise of graphene. Bi_2Se_3, Bi_2Te_3, and Sb_2Te_3 have been confirmed to have saturable absorption similar to that of graphene and have been applied in optical fiber and solid state laser to realize the short pulse output [18]. A possible route in the future is to combine graphene and two-dimensional materials to build heterostructures with complementary advantages, so as to obtain a better pulse output.

References

1. Bao, Q., Zhang, H., Wang, Y., Ni, Z., Yan, Y., Shen, Z. X., Loh, K. P., and Tang, D. Y. (2009). Atomic-layer graphene as a saturable absorber for ultrafast pulsed lasers *Advanced Functional Materials*, **19**, pp. 3077–3083.
2. Song, Y.-W., Jang, S.-Y., Han, W.-S., and Bae, M.-K. (2010). Graphene mode-lockers for fiber lasers functioned with evanescent field interaction, *Applied Physics Letters*, **96**, pp. 051122.
3. Choi, S. Y., Cho, D. K., Song, Y.-W., Oh, K., Kim, K., Rotermund, F., and Yeom, D.-I. (2012). Graphene-filled hollow optical fiber saturable absorber for efficient soliton fiber laser mode-locking, *Optics Express*, **20**, pp. 5652–5657.
4. Lin, Y.-H., Yang, C.-Y., Liou, J.-H., Yu, C.-P., and Lin, G.-R. (2013). Using graphene nano-particle embedded in photonic crystal fiber for evanescent wave mode-locking of fiber laser, *Optics Express*, **21**, pp. 16763–16776.
5. Sun, Z., Popa, D., Hasan, T., Torrisi, F., Wang, F., Kelleher, E. J., Travers, J. C., Nicolosi, V., and Ferrari, A. C. (2010). A stable, wideband tunable, near transform-limited, graphene-mode-locked, ultrafast laser, *Nano Research*, **3**, pp. 653–660.
6. Zhang, H., Tang, D. Y., Zhao, L. M., Bao, Q. L., Loh, K. P., Lin, B., and Tjin, S. C. (2010). Compact graphene mode-locked wavelength-tunable erbium-doped fiber lasers: From all anomalous dispersion to all normal dispersion, *Laser Physics Letters*, **7**, pp. 591–596.
7. Bao, Q., Zhang, H., Yang, J. X., Wang, S., Tang, D. Y., Jose, R., Ramakrishna, S., Lim, C. T., and Loh, K. P. (2010). Graphene–polymer nanofiber membrane for ultrafast photonics, *Advanced Functional Materials*, **20**, pp. 782–791.
8. Cunning, B., Brown, C., and Kielpinski, D. (2011). Low-loss flake-graphene saturable absorber mirror for laser mode-locking at sub-200-fs pulse duration, *Applied Physics Letters*, **99**, pp. 261109.
9. Liu, C., Ye, C., Luo, Z., Cheng, H., Wu, D., Zheng, Y., Liu, Z., and Qu, B. (2013). High-energy passively Q-switched 2 μm Tm^{3+}-doped double-clad fiber laser using graphene-oxide-deposited fiber taper, *Optics Express*, **21**, pp. 204–209.
10. Tan, W., Su, C., Knize, R., Xie, G., Li, L., and Tang, D. (2010). Mode locking of ceramic Nd:yttrium aluminum garnet with graphene as a saturable absorber, *Applied Physics Letters*, **96**, pp. 031106.

11. Xu, J.-L., Li, X.-L., Wu, Y.-Z., Hao, X.-P., He, J.-L., and Yang, K.-J. (2011). Graphene saturable absorber mirror for ultra-fast-pulse solid-state laser, *Optics Letters*, **36**, pp. 1948–1950.
12. Cho, W. B., Kim, J. W., Lee, H. W., Bae, S., Hong, B. H., Choi, S. Y., Baek, I. H., Kim, K., Yeom, D.-I., and Rotermund, F. (2011). High-quality, large-area monolayer graphene for efficient bulk laser mode-locking near 1.25 μm *Optics Letters*, **36**, pp. 4089–4091.
13. Liu, M., Yin, X., Ulin-Avila, E., Geng, B., Zentgraf, T., Ju, L., Wang, F., and Zhang, X. (2011). A graphene-based broadband optical modulator, *Nature*, **474**, pp. 64–67.
14. In Hyung, B., Hwang Woon, L., Sukang, B., Byung Hee, H., Yeong Hwan, A., Dong-Il, Y., and Fabian, R. (2012). Efficient mode-locking of sub-70-fs Ti:sapphire laser by graphene saturable absorber, *Applied Physics Express*, **5**, pp. 032701.
15. Cafiso, S. D. D. D., Ugolotti, E., Schmidt, A., Petrov, V., Griebner, U., Agnesi, A., Cho, W. B., Jung, B. H., Rotermund, F., Bae, S., Hong, B. H., Reali, G., and Pirzio, F. (2013). Sub-100-fs Cr: YAG laser mode-locked by monolayer graphene saturable absorber, *Optics Letters*, **38**, pp. 1745–1747.
16. Yu, W. J., Liu, Y., Zhou, H., Yin, A., Li, Z., Huang, Y., and Duan, X. (2013). Highly efficient gate-tunable photocurrent generation in vertical heterostructures of layered materials, *Nature Nanotechnology*, **8**, pp. 952–958.
17. Li, X.-L., Xu, J.-L., Wu, Y.-Z., He, J.-L., and Hao, X.-P. (2011). Large energy laser pulses with high repetition rate by graphene Q-switched solid-state laser, *Optics express*, **19**, pp. 9950–9955.
18. Zhao, C., Zhang, H., Qi, X., Chen, Y., Wang, Z., Wen, S., and Tang, D. (2012). Ultrashort pulse generation by a topological insulator based saturable absorber, *Applied Physics Letters*, **101**, pp. 211106.

Chapter 3

Graphene-Based Optical Modulators

Sheng Gan,[a] **Yupeng Zhang,**[b] **and Qiaoliang Bao**[a,b]

[a]*Institute of Functional Nano & Soft Materials (FUNSOM), Soochow University, 199 Ren-Ai Road, Suzhou Industrial Park, Suzhou Jiangsu 215123, China*
[b]*Department of Materials Science and Engineering, Monash University, 22 Alliance Lane, Building 69, Clayton, 3800, Victoria, Australia*
qiaoliang.bao@gmail.com

3.1 Introduction

Optical technology is poised to revolutionize short-reach interconnects. Integrated optical modulators with high bandwidth, small footprint, and broadband optical spectrum are a critical part of optical circuits. Therefore, semiconductor optical modulators have been heavily researched in the past few decades. However, it is confronted with a lot of intrinsic bottlenecks. Graphene has been hailed as a wonderful material in electronics and photonics since its discovery in 2004. Here we present some latest developments in graphene-based optical modulators.

It is well known that there are many characteristic parameters in optical signals, such as intensity, amplitude, frequency, phase, and polarization. The so-called optical modulation is to change

Graphene Photonics, Optoelectronics, and Plasmonics
Edited by Qiaoliang Bao, Hui Ying Hoh, and Yupeng Zhang

ISBN 978-981-4745-60-4 (Hardcover), 978-1-315-19667-1 (eBook)
www.panstanford.com

one or more of these parameters and encode them with various external forms of energy simultaneously. In addition, several figures of merit are used to characterize such an effect involving modulation depth, bandwidth, insertion loss, footprint, and power consumption. Optical modulation, achieved through electro-optics, thermo-optics, and acousto-optics, is a process of change in a material's effective refractive index, which accounts for the behavior of light propagation. Furthermore, electro-optic modulation, through which the effective linear susceptibility can be adjusted by applying an electric field across graphene, thus causing changes in both the real and imaginary refractive indices, has attracted great interest due to its high bandwidth. The change in the real part of the refractive index concerns electro-refraction, whereas the change in the imaginary counterpart concerns electro-absorption. At present, the active layer material of the modulator mainly consists of Si, III-IV compounds and $LiNbO_3$, which rely on many different effects, including the Pockels effect, the Kerr effect, the Franz–Keldysh effect, and the quantum-confined Stark effect (QCSE). Depending on which effect is manifesting, the modulators can be classified as either electro-refractive or electro-absorptive. However, these modulators suffer from large footprint, difficult integration with CMOS technique, and narrow bandwidth [1–4].

3.2 Graphene Strip-Waveguide Optical Modulators

Graphene has attracted growing attention due to its excellent optical and electrical properties since its discovery in 2004 [5–10]. Moreover, the broad optical absorption can be controlled through electrical gating [11, 12]: By shifting the electronic Fermi level, one can control optical transitions. Liu et al. fabricated a single-layer graphene optical modulator for the first time in 2011 [13]. The device has many advantages, such as a broad optical spectrum (1.35–1.6 μm), large modulation depth (0.1 dB/μm), and small footprint (25 μm^2). Graphene has a great potential to be used for optical modulators owing to its ultrahigh carrier mobility [14, 15], and compatibility with CMOS techniques make it a bright

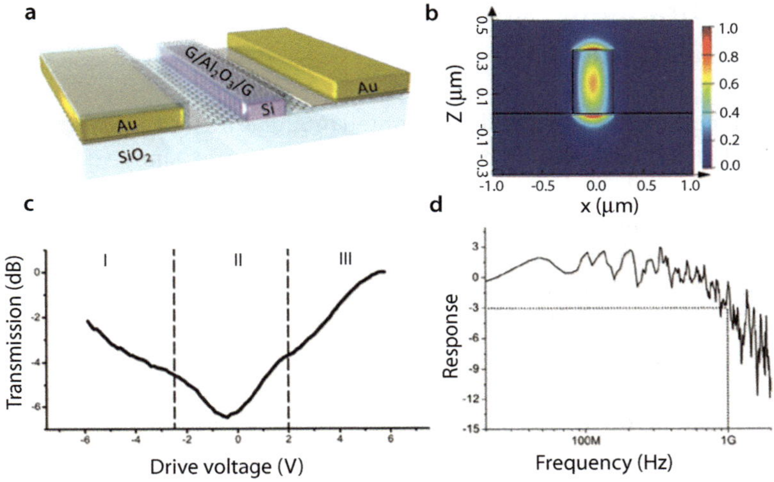

Figure 3.1 (a) Three-dimensional schematic illustration of the device. (b) Cross section of the device, with an overlay of the optical mode plot, calculated by finite element simulation (FDTD solution). (c) Static response of a double-layer graphene modulator. (d) Dynamic response of the device. Reprinted with permission from Ref. 19, Copyright 2012, American Chemical Society.

future in optical modulator [16–18]. On the basis of previous fabrications, the authors fabricated another device based on double-layer graphene separated by a thin layer of Al_2O_3 [19]. Voltage can be applied to the graphene to form a modulating electric field so that Fermi level can be changed. Optical absorption happens only when the Fermi level falls between the thresholds of $\pm\, h\nu_0/2$; otherwise graphene is transparent. This kind of configuration avoids disadvantages inherited from silicon photonics, such as high insertion loss and limited carrier mobility. Meanwhile, extinction ratio improves significantly due to stronger interaction between the light field and double-layer graphene. The structure of the electro-absorption modulator and its optical mode plot are schematically illustrated in Figs. 3.1a,b. Figures 3.1c,d show static and dynamic responses of the device from which we observe that modulation depth and bandwidth are able to reach 0.16 dB/μm and 1 GHz, respectively. This is comparable to the traditional optical modulators made of Si and III-V compounds [20, 21].

3.3 Graphene-Slot Waveguide Optical Modulators

Currently, these waveguide-integrated modulators face an important challenge that overlay between graphene and optical mode dose not reach the strongest. To this end, Kim et al. proposed a model of ridge waveguide modulator in 2011, as illustrated in Fig. 3.2a [22]. The configuration shown in Fig. 3.2b with a layer of p-Si deposited on the top of Si of modulators discussed earlier is capable of confining light field to graphene in a large degree. Besides, Al_2O_3, which deteriorates the mobility of graphene, is replaced by hBN with a low permitivity so that RC constant reduces significantly. Consequently, modulation depth and speed improve dramatically.

Gosciniak et al. continued to research a similar modulator, as illustrated in Figs. 3.2c,d [23]. It is claimed that the highest modulation depth and speed of 5.05 dB/μm and 510 GHz, respectively, are achievable with proper structure according to calculation. At the same time, the corresponding bit energy is 0.96 fJ/bit and optical bandwidth is 14 THz. In 2012, Lu et al. put forward two new types of slot waveguide modulators (Si and plasmonic waveguide), given in Figs. 3.2e–g [24].

Mechanisms of the new type of modulators are little different: When chemical potential μ_c (namely E_F) is 0 eV, interband transition occurs and the modulator is ON because the power absorbed per unit area (P_d) is minimum with effective dielectric constant maximum and light field in graphene weakest; while μ_c is 0.515 eV, intraband transition happens now and P_d becomes maximum resulting in OFF state with minimum permittivity and light field. A further study of these two devices shown in Figs. 3.2f,h was carried out by FDTD solution. The results show that the modulation depth of the modulator based on a silicon waveguide can reach 4.40 dB/μm, that is to say 3 dB is achieved with active layer length of only 681 nm. Moreover, the intrinsic ability to confine optical mode of plasmonic waveguide modulator makes its largest modulation depth 7.14 dB/μm with 420 nm length to realize 3 dB. The corresponding bit energy is low (0.12–0.13 pJ/bit) and thermal stability is high simultaneously based on the calculations. Intraband transition, whose relaxation time is shorter than that of interband

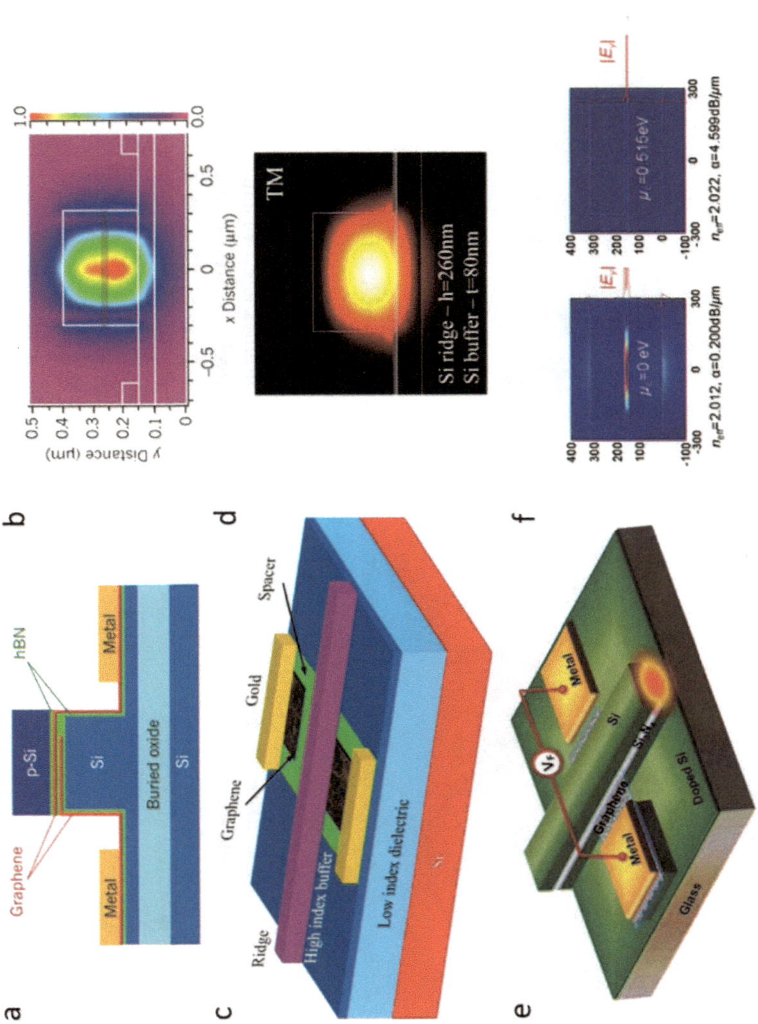

Figure 3.2 *(Continued)*

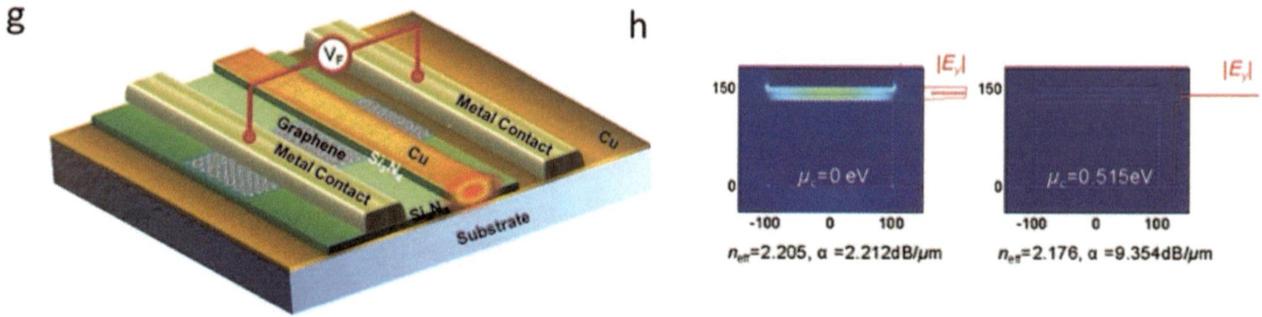

Figure 3.2 (a) A ridge-type modulator. (b) Transverse electric mode profile at a wavelength of 1.55 μm is shown for the ridge-type modulator. Reprinted with permission from Ref. 22, Copyright 2011, Nature Publishing Group. (c) The illustration of the rib waveguide structure. (d) Field distribution plot of the magnitude of the power flow for considered structure with $\mu_c = 0.512$ eV. Reprinted with permission from Ref. 23, Copyright 2013, Nature Publishing Group. (e) Schematic of a graphene electro-optic (EO) modulator based on a silicon waveguide. (f) The transverse electric field profiles, effective indices, and propagation loss for silicon graphene-slot waveguide when $\mu_c = 0$ eV and $\mu_c = 0.515$ eV, respectively. (g) Schematic of a graphene EO modulator based on a metal strip plasmonic waveguide. (h) The transverse electric field profiles, effective indices, and propagation loss for silicon graphene-slot waveguide when $\mu_c = 0$ eV and $\mu_c = 0.515$ eV, respectively. Reprinted with permission from Ref. 24, Copyright 2012, Optical Society of America.

transition, involves in the optical switch, so the device can work in ultrahigh frequency theoretically.

3.4 Optical Modulators with MZI and Ring Structures

There are two features of small footprint and superior extinction ratio in the modulator based on ring resonator [4], which caused a significant amount of attention since it was fabricated in 2005. In 2011, Bao et al. were the first to come up with such a modulator with a single-layer graphene, as illustrated in Fig. 3.3a [25]. Midrio et al. went on to further research the aforementioned device with commercial codes CST and COMSOL, as given in Figs. 3.3b–d [26].

The operation principle is as follows: We can modulate the absorption coefficient of graphene by altering its Fermi level with voltage applied to the double-layer graphene separated by Al_2O_3. The switch between ON (non-critical coupling) and OFF (critical coupling) states is achievable through variation of the transmission coefficient of ring resonator. Simulation results show that its extinction ratio, voltage required for charging, and bit energy can be 44 dB, 1.2 V, and 10∼30 fJ/bit, respectively, with a theoretical 3 dB bandwidth of 100 GHz. Besides, the Mach–Zehnder modulator (MZM) is also a popular design because of its superior optical spectrum and higher tolerance for fabrication and temperature.

In 2012, Grigorenko et al. proposed the first model of graphene plasmonic MZM, shown in Fig. 3.3e [27]. Figure 3.3f shows another kind of MZM put forward by Hao et al. in September 2013 [28]. Eight layers of graphene separated by Si were mechanically transferred onto the waveguide sequentially forming a graphene-embedded structure resulting in significant enhancement of the electro-refractive effect. When two identical voltages are applied to the double arms, the output light interferes constructively, resulting in the ON state; while the difference between the two voltages is half-wave voltage V_{Π}, the output light interferes destructively leading to the OFF state. Additionally, its modulation efficiency and extinction ratio can be 20 V·μm and 35 dB, respectively, calculated by COMSOL.

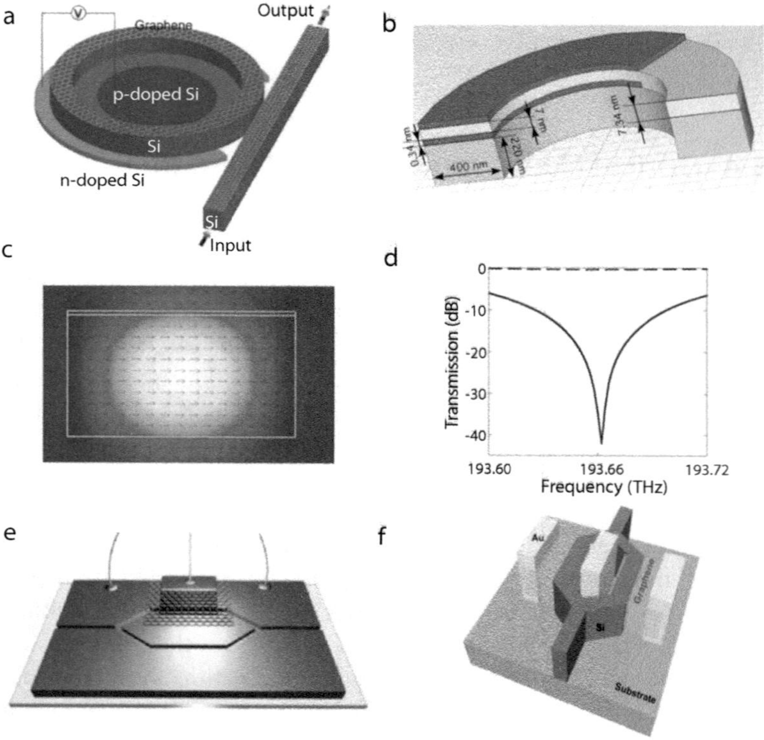

Figure 3.3 (a) Proposed model of graphene-based ring resonator. Reprinted with permission from Ref. 25, Copyright 2012, American Chemical Society. (b) Schematic diagram of the ring waveguide. (c) Power density and lines of force of the electric field of the fundamental TE mode. (d) Transmission in a critically coupled ring with gap $g = 100$ nm, average radius equal to 5 μm for OFF (solid line) and ON (dashed curve). Reprinted with permission from Ref. 26, Copyright 2012, Optical Society of America. (e) Hypothetical graphene-based active plasmonic modulator. Reprinted with permission from Ref. 27, Copyright 2012, Nature Publishing Group. (f) Schematic illustration for top view of the Mach–Zehnder modulator configuration. Reprinted from Ref. 28 with the permission of AIP Publishing.

Apart from these proposed simulation models, Youngblood et al. fabricated the multifunctional graphene optical modulator and photodetector integrated on silicon waveguides in 2014 (Fig. 3.4) [29]. The graphene device is tunable with a gate made from another layer

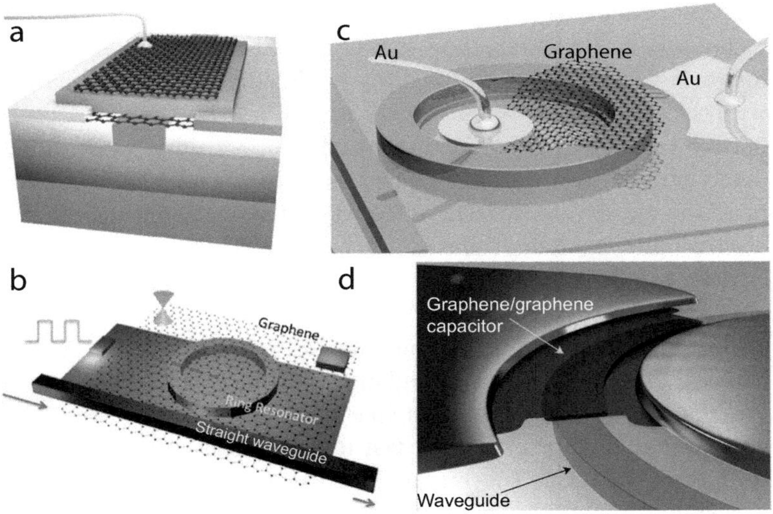

Figure 3.4 (a) Schematic illustration of the dual-layer graphene modulator/detector integrated on a planarized waveguide. Reprinted with permission from Ref. 29, Copyright 2014, American Chemical Society. (b) Schematic diagram of a graphene-modulated silicon microring modulator. Reprinted with permission from Ref. 30, Copyright 2014, American Chemical Society. (c) Three-dimensional schematic diagram of the designed graphene–silicon waveguide. Reprinted with permission from Ref. 31, Copyright 2015, American Chemical Society. (d) Schematic illustration of the modulator consisting of a graphene/graphene capacitor integrated along a ring resonator. Reprinted with permission from Ref. 32, Copyright 2015, Nature Publishing Group.

of graphene to achieve near-infrared photodetection responsivity of 57 mA/W and modulation depth of 64%. This novel multifunctional device may lead to many unprecedented optoelectronic applications. In the same year, Qiu et al. demonstrated a $\sim$40% amplitude modulation of 1.55 μm radiation with a gated single-layer graphene coupled with a silicon microring resonator [30]. Both the quality factor and resonance wavelength of the silicon microring resonator were strongly modulated through gate tuning of the Fermi level in graphene.

These results promise an efficient electro-optic modulator, ideal for applications in large-scale on-chip optical interconnects that are compatible with complementary metal-oxide-semiconductor technology. In May 2015, Ding et al. demonstrated graphene–silicon microring devices, showing a high modulation depth of 12.5 dB with a relatively low bias voltage of 8.8 V [31]. On–off electro-optical switching with an extinction ratio of 3.8 dB is successfully demonstrated by applying a square waveform with a 4 V peak-to-peak voltage. In August 2015, Phare et al. demonstrated a graphene electro-optic modulator based on resonator loss modulation at critical coupling, which shows drastically increased speed and efficiency [32].

In our group, our device operates with a 30 GHz bandwidth and with a state-of-the-art modulation efficiency of 15 dB per 10 V. We also showed the first high-speed large-signal operation in a graphene modulator, paving the way for fast digital communications using this platform. The modulator uniquely uses silicon nitride waveguides, an otherwise completely passive material platform, with promising applications for ultra-low-loss broadband structures and nonlinear optics.

3.5 Optical Modulators with Vertical Structure

The aforementioned modulators are all of small dimension on the order of micrometer owning fast modulation speed, which is suitable for optical interconnects. However, they are useless for applications where there is a need for free space of low insertion loss, such as amplitude modulation of high quality laser resonators. To solve this problem, Lee et al. fabricated in 2012 the first reflective plane-incidence modulator based on a single-layer graphene, as presented in Fig. 3.5a [33]. The underlying Ag mirror acts as both a back gate and a reflector. Figure 3.5b illustrates the dynamic response of the device from which we can know that the 3 dB bandwidth is 154 MHz. Moreover, the feature of the active layer area being above 7850 μm^2 makes it a good candidate for lasers and active interferometers.

In 2013, Polat et al. designed another structure, as shown in Fig. 3.5c [34]. Double layers of graphene were mechanically transferred onto two quartz substrates, respectively, and then placed

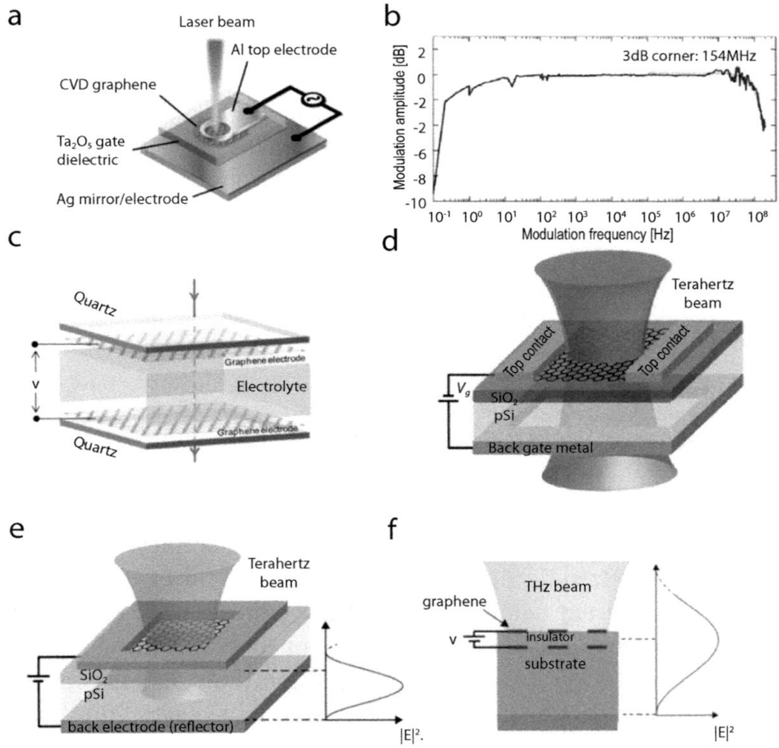

Figure 3.5 (a) Sketch of the device structure. (b) Frequency response of the graphene modulator. Figures (a) and (b) reprinted with permission from Ref. 33, Copyright 2012, The Optical Society. (c) Schematic exploded view of the optically transparent double-layer capacitor. Reprinted with permission from Ref. 34, Copyright 2013, American Chemical Society. (d) Schematic diagram of the graphene THz transmission modulator. Reprinted with permission from Ref. 35, Copyright 2012, Nature Publishing Group. (e) Schematic diagram of the graphene THz reflection modulator. Reprinted with permission from Ref. 36, Copyright 2012, American Chemical Society. (f) Schematic diagram of the device, and electric field distribution when the substrate thickness and plasmonic resonance are matched to an odd multiple of a quarter wavelength of the THz wave. Reprinted from Ref. 37 with the permission of AIP Publishing.

together face to face separated by an electrolyte, forming a graphene supercapacitor. Optical modulation can be achieved due to the adjustment of the Fermi level of graphene which a voltage is applied to. Multilayer graphene and reflective structure able to enhance the interaction between graphene and light field are employed to avoid inferior modulation depth using a single-layer graphene. Finally, ultrahigh optical bandwidth (450 nm to 2 μm) and modulation efficiency (35%) are achievable.

Apart from the modulation based on optical interband transition, the modulation of optical intraband transition is also possible [35, 36]. Figure 3.5d presents a broadband graphene terahertz (THz) transmission modulator designed by Rodriguez et al. in 2012 [35]. Fermi level locates at Dirac point with no gate voltage (V_g) so that the density of state equals to 0 and no photon is absorbed, thus bringing about the ON state. It turns to OFF when a V_g is applied. Its modulation efficiency and speed are 15% and 20 KHz, respectively, referred from the measurement. In the same year, they fabricated a graphene THz reflection modulator, as shown in Fig. 3.5e [36]. There are different points between the two modulators: Back gate of the latter device also serves as a reflector so that the incident and reflected light interfere constructively when the substrate optical thickness is an odd multiple of the THz wavelength; thus, the field intensity in graphene is at maxima. Its measured modulation efficiency is 64%, which improves significantly. And the associated modulation speed around 4 KHz with an insertion loss slightly below 2 dB is observed.

Recently, another model of a THz reflection modulator employing graphene plasmonic structure was put forward by the same group, as illustrated in Fig. 3.5f [37]. From this diagram, we can see that graphene is patterned into micro-ribbons. Thus, taking advantage of the plasmonic effects, as well as electric field enhancement in the active regions of the device by a back reflector, it is shown that a very high modulation depth at any frequency in the THz range can be achieved. A study shows that there is such a relationship among graphene plasmon resonance frequency ω_p, ribbon width w, and carrier concentration n: $\omega_p \propto w^{-1/2} \cdot n^{1/4}$ [38]. Therefore, when ω_p is equal to the optical wavelength with an appropriate V_g applied, it causes plasmonic effect that enhances the absorption coefficient of graphene resulting in the OFF state; it turns to ON as long as $V_g = 0$ with no photon absorption. It is theoretically calculated that a device with $w = 2$ μm and filling factor 0.5 can meet an unexpected performance of modulation efficiency reaching 70% with insertion loss of 0.8 dB.

3.6 Conclusion

Until now, much progress has been made in graphene-based optical modulators, but it does not match optical interconnects so well, which need high bandwidth, large modulation depth, and low insertion loss, because of some shortcomings. The main obstacle is the limitation caused by the RC constant, which mainly comes from contact resistance between graphene and electrodes. Therefore, reducing the contact resistance is a critical research direction in the field of future modulators.

References

1. Kuo, Y. H., Lee, Y. K., Ge, Y. S., Ren, S., Roth, J. E., Kamins, T. I., Miller, D. A. B., and Harris, J. S. (2005). Strong quantum-confined Stark effect in germanium quantum-well structures on silicon. *Nature*, **437**, pp. 1334–1336.
2. Liu, A. S., Jones, R., Liao, L., Samara-Rubio, D., Rubin, D., Cohen, O., Nicolaescu, R., and Paniccia, M. (2004). A high-speed silicon optical modulator based on a metal-oxide-semiconductor capacitor. *Nature*, **427**, pp. 615–618.
3. Miller, D. A. B., Chemla, D. S., Damen, T. C., Gossard, A. C., Wiegmann, W., Wood, T. H., and Burrus, C. A. (1984). Band-edge electroabsorption in quantum well structures: The quantum-confined Stark-effect. *Phys. Rev. Lett.*, **53**, pp. 2173–2176.
4. Xu, Q. F., Schmidt, B., Pradhan, S., and Lipson, M. (2005). Micrometre-scale silicon electro-optic modulator. *Nature*, **435**, pp. 325–327.
5. Avouris, P., Chen, Z. H., and Perebeinos, V. (2007). Carbon-based electronics. *Nat. Nanotechnol.*, **2**, pp. 605–615.
6. Liao, L., Lin, Y. C., Bao, M. Q., Cheng, R., Bai, J. W., Liu, Y. A., Qu, Y. Q., Wang, K. L., Huang, Y., and Duan, X. F. (2010). High-speed graphene transistors with a self-aligned nanowire gate. *Nature*, **467**, pp. 305–308.

7. Novoselov, K. S., Geim, A. K., Morozov, S. V., Jiang, D., Zhang, Y., Dubonos, S. V., Grigorieva, I. V., and Firsov, A. A. (2004). Electric field effect in atomically thin carbon films. *Science*, **306**, pp. 666–669.
8. Schwierz, F. (2010). Graphene transistors. *Nat. Nanotechnol.*, **5**, pp. 487–496.
9. Geim, A. K., and Novoselov, K. S. (2007). The rise of graphene. *Nat. Mater.*, **6**, pp. 183–191.
10. Bonaccorso, F., Sun, Z., Hasan, T., and Ferrari, A. C. (2010). Graphene photonics and optoelectronics. *Nat. Photonics*, **4**, pp. 611–622.
11. Li, Z. Q., Henriksen, E. A., Jiang, Z., Hao, Z., Martin, M. C., Kim, P., Stormer, H. L., and Basov, D. N. (2008). Dirac charge dynamics in graphene by infrared spectroscopy. *Nat. Phys.*, **4**, pp. 532–535.
12. Wang, F., Zhang, Y. B., Tian, C. S., Girit, C., Zettl, A., Crommie, M., and Shen, Y. R. (2008). Gate-variable optical transitions in graphene. *Science*, **320**, pp. 206–209.
13. Liu, M., Yin, X. B., Ulin-Avila, E., Geng, B. S., Zentgraf, T., Ju, L., Wang, F., and Zhang, X. (2011). A graphene-based broadband optical modulator. *Nature*, **474**, pp. 64–67.
14. Bolotin, K. I., Sikes, K. J., Jiang, Z., Klima, M., Fudenberg, G., Hone, J., Kim, P., and Stormer, H. L. (2008). Ultrahigh electron mobility in suspended graphene. *Solid State Commun.*, **146**, pp. 351–355.
15. Du, X., Skachko, I., Barker, A., and Andrei, E. Y. (2008). Approaching ballistic transport in suspended graphene. *Nat. Nanotechnol.*, **3**, pp. 491–495.
16. Kim, K. S., Zhao, Y., Jang, H., Lee, S. Y., Kim, J. M., Kim, K. S., Ahn, J. H., Kim, P., Choi, J. Y., and Hong, B. H. (2009). Large-scale pattern growth of graphene films for stretchable transparent electrodes. *Nature*, **457**, pp. 706–710.
17. Reina, A., Jia, X. T., Ho, J., Nezich, D., Son, H. B., Bulovic, V., Dresselhaus, M. S., and Kong, J. (2009). Layer area, few-layer graphene films on arbitrary substrates by chemical vapor deposition. *Nano Lett.*, **9**, pp. 3087–3087.
18. Bae, S., Kim, H., Lee, Y., Xu, X. F., Park, J. S., Zheng, Y., Balakrishnan, J., Lei, T., Kim, H. R., Song, Y. I., Kim, Y. J., Kim, K. S., Ozyilmaz, B., Ahn, J. H., Hong, B. H., and Iijima, S. (2010). Roll-to-roll production of 30-inch graphene films for transparent electrodes. *Nat. Nanotechnol.*, **5**, pp. 574–578.
19. Liu, M., Yin, X. B., and Zhang, X. (2012). Double-layer graphene optical modulator. *Nano Lett.*, **12**, pp. 1482–1485.

20. Liu, J., Beals, M., Pomerene, A., Bernardis, S., Sun, R., Cheng, J., Kimerling, L. C., and Michel, J. (2008). Waveguide-integrated, ultralow-energy GeSi electro-absorption modulators. *Nat. Photonics*, **2**, pp. 433–437.
21. Reed, G. T., Mashanovich, G., Gardes, F. Y., and Thomson, D. J. (2010). Silicon optical modulators. *Nat. Photonics*, **4**, pp. 518–526.
22. Kim, K., Choi, J. Y., Kim, T., Cho, S. H., and Chung, H. J. (2011). A role for graphene in silicon-based semiconductor devices. *Nature*, **479**, pp. 338–344.
23. Gosciniak, J., and Tan, D. T. H. (2013). Theoretical investigation of graphene-based photonic modulators. *Sci Rep-Uk*, 3, 1897.
24. Lu, Z. L., and Zhao, W. S. (2012). Nanoscale electro-optic modulators based on graphene-slot waveguides. *J. Opt. Soc. Am. B.*, **29**, pp. 1490–1496.
25. Bao, Q. L., and Loh, K. P. (2012). Graphene photonics, plasmonics, and broadband optoelectronic devices. *ACS Nano*, **6**, pp. 3677–3694.
26. Midrio, M., Boscolo, S., Moresco, M., Romagnoli, M., De Angelis, C., Locatelli, A., and Capobianco, A. D. (2012). Graphene-assisted critically-coupled optical ring modulator. *Opt. Express*, **20**, pp. 23144–23155.
27. Grigorenko, A. N., Polini, M., and Novoselov, K. S. (2012). Graphene plasmonics. *Nat. Photonics*, **6**, pp. 749–758.
28. Hao, R., Du, W., Chen, H. S., Jin, X. F., Yang, L. Z., and Li, E. P. (2013). Ultra-compact optical modulator by graphene induced electro-refraction effect. *Appl. Phys. Lett.*, **103**, 061116.
29. Youngblood, N., Anugrah, Y., Ma, R., Koester, S. J., and Li, M. (2014). Multi-functional graphene optical modulator and photodetector integrated on silicon waveguides. *Nano Lett.*, **14**, pp. 2741–2746.
30. Qiu, C., Gao, W., Vajtai, R., Ajayan, P. M., Kono, J., and Xu, Q. (2014). Efficient modulation of 1.55 μm radiation with gated graphene on a silicon microring resonator. *Nano Lett.*, **14**, pp. 6811–6815.
31. Ding, Y. H., Zhu, X. L., Xiao, S. S., Hu, H., Frandsen, L. H., Mortensen, N. A., and Yvind, K. (2015). Effective electro-optical modulation with high extinction ratio by a graphene-silicon microring resonator. *Nano Lett.*, **15**, pp. 4393–4400.
32. Phare, C. T., Lee, Y. H. D., Cardenas, J., and Lipson, M. (2015). Graphene electro-optic modulator with 30 GHz bandwidth. *Nat. Photonics*, **9**, pp. 511–514.

33. Lee, C. C., Suzuki, S., Xie, W., and Schibli, T. R. (2012). Broadband graphene electro-optic modulators with sub-wavelength thickness. *Opt. Express*, **20**, pp. 5264–5269.
34. Polat, E. O., and Kocabas, C. (2013). Broadband optical modulators based on graphene supercapacitors. *Nano Lett.*, **13**, pp. 5851–5857.
35. Sensale-Rodriguez, B., Yan, R. S., Kelly, M. M., Fang, T., Tahy, K., Hwang, W. S., Jena, D., Liu, L., and Xing, H. G. (2012). Broadband graphene terahertz modulators enabled by intraband transitions. *Nat. Commun.*, **3**, 780.
36. Sensale-Rodriguez, B., Yan, R. S., Rafique, S., Zhu, M. D., Li, W., Liang, X. L., Gundlach, D., Protasenko, V., Kelly, M. M., Jena, D., Liu, L., and Xing, H. G. (2012). Extraordinary control of terahertz beam reflectance in graphene electro-absorption modulators. *Nano Lett.*, **12**, pp. 4518–4522.
37. Sensale-Rodriguez, B., Yan, R. S., Zhu, M. D., Jena, D., Liu, L., and Xing, H. G. (2012). Efficient terahertz electro-absorption modulation employing graphene plasmonic structures. *Appl. Phys. Lett.*, **101**, 261115.
38. Ju, L., Geng, B. S., Horng, J., Girit, C., Martin, M., Hao, Z., Bechtel, H. A., Liang, X. G., Zettl, A., Shen, Y. R., and Wang, F. (2011). Graphene plasmonics for tunable terahertz metamaterials. *Nat. Nanotechnol.*, **6**, pp. 630–634.

Chapter 4

Graphene-Based Optical Polarizers

Ziyu Wang,[a] **Zai-Quan Xu,**[a] **Yupeng Zhang,**[a]
and Qiaoliang Bao,[a,b]

[a] *Department of Materials Science and Engineering, Monash University, 22 Alliance Lane, Building 69, Clayton, 3800, Victoria, Australia*
[b] *Institute of Functional Nano & Soft Materials (FUNSOM), Soochow University, 199 Ren-Ai Road, Suzhou Industrial Park, Suzhou Jiangsu 215123, China*
qiaoliang.bao@gmail.com

4.1 Introduction

Polarization is a common property of electromagnetic waves. If the electric vectors are all along the same plane, the light is polarized; otherwise, the light is unpolarized. A polarizer is an optical filter that passes light of a specific polarization and blocks waves of other polarizations. It can convert a beam of light of unpolarized or mixed polarization into a beam with well-defined polarization, polarized light. All fiber in-line polarizer is one of the most important components in fiber-optic communication and sensor systems. The in-line devices are constructed by first polishing a short section of the lateral surface of the cladding to within the evanescent field around the fiber core, followed with cover with crystals [1, 2], thin metal films [3–5], or graphene [6, 7]. In a typical device, unpolarized

Graphene Photonics, Optoelectronics, and Plasmonics
Edited by Qiaoliang Bao, Hui Ying Hoh, and Yupeng Zhang

ISBN 978-981-4745-60-4 (Hardcover), 978-1-315-19667-1 (eBook)
www.panstanford.com

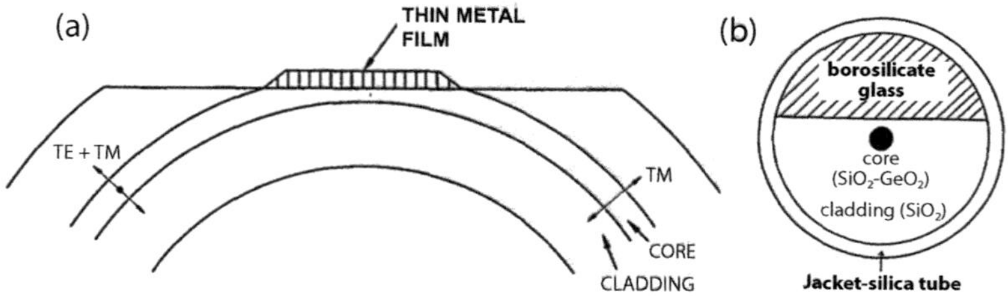

Figure 4.1 (a) Schematic diagram of a metal-clad cutoff fiber-optic polarizer. Reprinted from Ref. 8 with permission from The Optical Society. (b) Fiber polarizer based on crystal with larger refractive index. Reprinted from Ref. 12 with permission from The Optical Society.

or mix polarization light is coupled into and guided by the fiber from one end. The evanescent field of the light guided by the fiber interacts with the selective material covered on the polished parts along the fiber. The desired polarized light remains unaffected by the overlay material and is still guided by the fiber, while the light with unwanted polarization interacts with atop material and is no longer guided by the fiber. Only light with the desired polarization is coupled out in this manner.

4.2 Conventional In-Line Polarizers

Thin-metal-film in-line polarizers were developed by W. Eickhoff in 1980 (Fig. 4.1) [3]. Au or Al was deposited on the polished section of the fiber. An extinction ratio of 40 dB was achieved. In the following decades, several metal materials, such as In, Ag, and a combination of MgF_2 and Al, and birefingent materials, such as $KB_5O_8{\cdot}4H_2O$ and polyvinyl carbazole (PVK) [1, 4, 8, 9], were used in the in-line fiber devices to achieve high extinction ratio with low loss. Recently, plasmonic nanostructures such as gold nano-gratings and 2D nanosheets have emerged as novel materials with the development of nanotechnology [10, 11]. Table 4.1 summarizes the materials, working wavelength, extinction ratio, and insertion loss.

Now we discuss the two working mechanisms. For fiber polarizer based on birefringent crystals, the materials are to fill the parts where cladding is removed. Normally, the refractive index of the materials used is greater than the effective index of the fiber. The

Table 4.1 Common materials and corresponding properties for in-line polarizers

Interaction Material	Wavelength (nm)	Out Mode	Extinction Ratio (dB)	Insertion Loss (dB)
Au (150 nm) [3]	400–800	TM	14	NA
Al (150 nm) [3]	400–800	TM	NA	NA
$KB_5O_8{\cdot}4H_2O$ [2]	1455–1465	TM	60	NA
Al [12]	1290	TM	22.2	4.2
Al [12]	1150	TM	36.6	10.5
Ag [8]	820	TM	50	6
Au [8]	820	TM	47	1
In [4]	NA	TM		
Al/MgF_2 [9]	600–1600	TM	80	0.5
Al/MgF_2 [19]	633	TM	63	

light guided within the fiber could excite a bulk wave in the crystal and the light escapes from the fiber. On the other hand, if the refractive index of the materials is less than the effective index of the waveguide, no bulk wave is excited and no light escapes from the fiber [2]. For metal- or graphene-based fiber polarizer, a fiber-to-surface plasmon mode coupling theory applies. A surface plasmon is an electromagnetic wave that propagates along the interface of two materials, one of which has a negative dielectric constant. Surface plasmons are collective oscillations of electrons that occur on the surfaces of solid materials that contain free electrons. The surface plasmon resonance depends on the coupling of light incident on a thin metal film into surface electron oscillations within the film.

4.3 Graphene-Based Polarizers

4.3.1 Optical Fiber Polarizers

Graphene has emerged as a single layer of carbon atoms arranged in a hexagonal layout since its first discovery in 2004. The electrons of graphene are confined in a 2D system; hence the electromagnetic wave propagating is governed by [13]:

$$1 + \frac{2\pi i \sigma(\omega)\sqrt{q^2 - \omega^2/c^2}}{\omega} = 0 \tag{4.1}$$

for TM mode, and by

$$1-\frac{2\pi i\sigma(\omega)}{\omega\sqrt{q^2-\omega^2/c^2}}=0 \tag{4.2}$$

for TE mode, where $\sigma(\omega)$ is the local dynamic conductivity of the 2D materials and c is the velocity of light. Here, $\sigma(\omega)=\sigma'+i\sigma''$, where σ' and σ'' are real functions of frequency. σ'' can be either positive or negative depending on the frequency and the electronic chemical potential.

As a consequence, graphene can selectively support either TM or TE electromagnetic modes. Based on this theory, it should be promising to develop a graphene/silica hybrid waveguide, which selects either a TM or TE mode, thus transforming unpolarized light into polarized light [14]. Recently, Bao et al., successfully coupled broadband light into graphene film by transferring graphene grown with chemical vapor deposition onto a side-polished optical fiber to form a graphene-to-fiber coupler, as shown in Fig. 4.2 [6]. As only pristine graphene is used, the TE surface wave is supported, whereas the TM wave is scattered via leaky mode. The TE mode polarizer

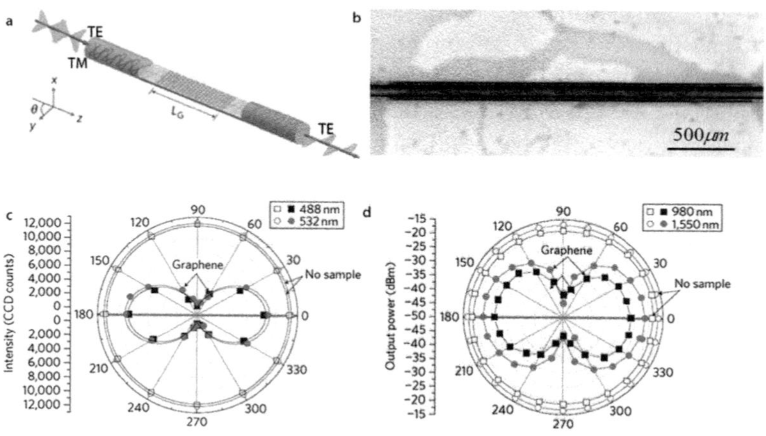

Figure 4.2 Graphene-based in-line polarizer and broadband polarization effect. (a) Schematic illustration of a fiber-to-graphene coupler on a side-polished fiber. (b) Optical image of side-polished fiber covered by a few layer of graphene films. (c) Polar image measured at 488 and 532 nm. (d) Polar image measured at 980 and 1550 nm. Reprinted with permission from Ref. 6, Copyright 2011, Nature Publishing Group.

works over a wide range from visible to near infrared and shows strong polarization effect within telecommunications bands with an extinction ratio of ~27 dB.

4.3.2 Optical Waveguide Polarizers

The graphene-based TE polarizer can be switched into the TM mode if the graphene film is sufficiently doped to shift to chemical potential to a value larger than half excitation energy. Inspired by this work, Kim et al. reported graphene-based waveguide polarizer concepts with averaged extinction ratio of 19 dB, which can support either TE or TM mode, as shown in Fig. 4.3 [16–18]. As mentioned earlier, the interaction between electromagnetic wave and graphene can be tuned by chemical doping, external electrical field, and magnetic field. Kim et al. also demonstrated a TE-pass graphene-based polarizer with an extinction ratio of 10 dB by transferring graphene strip onto the waveguide core, as shown in Fig. 4.3a. However, if the air-cladding is replaced by a polymer-cladding with the same refractive index, the TE-pass polarizer is switched to TM-pass polarizer with an extinction ratio of 19 dB at a wavelength of 1310 nm. This transition occurs because the electrical properties of graphene change on contact with polymer-cladding.

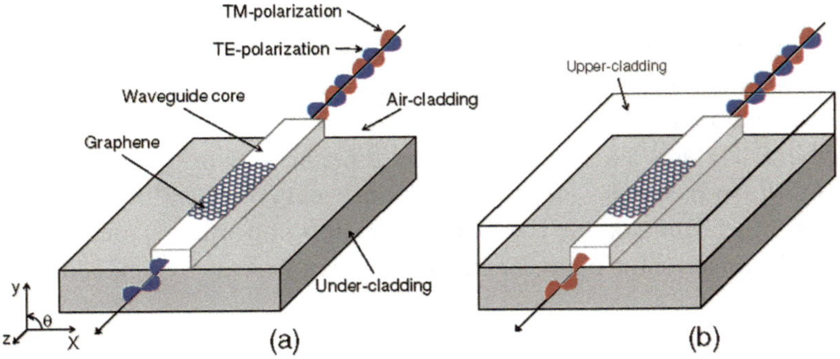

Figure 4.3 Schematic illustration of graphene-based waveguide polarizers. (a) Graphene-based waveguide polarizer with air-cladding that supports TE mode. (b) Graphene-based waveguide polarizer with polymer-cladding that supports TM mode. Reprinted from Ref. 8 with permission from The Optical Society.

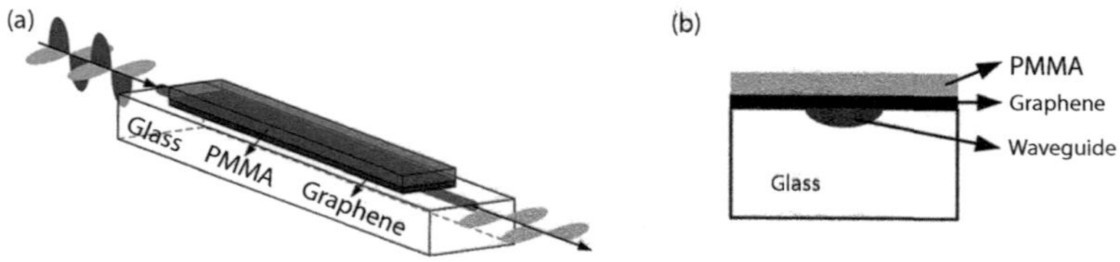

Figure 4.4 (a) Schematic illustration of a graphene/glass hybrid waveguide polarizer. (b) Schematic cross section of such a polarizer. Reprinted with permission from Ref. 18, Copyright 2015, IEEE.

Additionally, in order to increase the efficiency of graphene-based waveguide polarizers, Pei et al. selected PMMA film as the upper cladding to reduce the doping level of the graphene, as shown in Fig. 4.4. This modification leads the extinction ratio of TE-pass waveguide-based polarizers to increase to about 28 dB at a wavelength of 1370 nm [18]. In addition to the advantages of simpler fabrication and lower cost compared to metal-clad polarizers, graphene is chemically tunable, which allows the fabrication of environment-sensitive polarizers in the future.

4.4 Conclusion

In-line polarizers have come a long way since the 1980s [2–4, 19]. Various materials for in-line polarizers have been realized, such as gold nanoparticles, polymer, and graphene deposited on D-shaped fibers. TM and TE mode polarizers have also been demonstrated. Graphene-based in-line polarizers with easy fabrication and low cost are promising for the next generation of products, and future works on dynamic mode-switching are expected.

References

1. Lee, S. G., Sokoloff, J. P., McGinnis, B. P., and Sasabe, H. (1997). Fabrication of a side-polished fiber polarizer with a birefringent polymer overlay, *Optics Letters*, **22**, pp. 606–608.

2. Bergh, R. A., Lefevre, H. C., and Shaw, H. J. (1980). Single-mode fiber-optic polarizer, *Optics Letters*, **5**, pp. 479–481.
3. Eickhoff, W. (1980). In-line fiber-optic polariser, *Electronics Letters*, **16**, pp. 762–763.
4. Dyott, R. B., Bello, J., and Handerek, V. A. (1986). Indium-coated D-shaped-fiber polarizer, *Optics Letters*, **12**, pp. 287–289.
5. Skovsen, E., Søndergaard, T., Lemke, C., Holmgaard, T., Leißner, T., Eriksen, R. L., Beermann, J., Bauer, M., Pedersen, K., and Bozhevolnyi, S. I. (2013). Plasmonic black gold based broadband polarizers for ultra-short laser pulses, *Applied Physics Letters*, **103**, pp. 211102.
6. Bao, Q., Zhang, H., Wang, B., Ni, Z., Lim, C. H. Y. X., Wang, Y., Tang, D. Y., and Loh, K. P. (2011). Broadband graphene polarizer, *Nature Photonics*, **5**, pp. 411–415.
7. Lim, W. H., Yap, Y. K., Chong, W. Y., Pua, C. H., Huang, N. M., De La Rue, R. M., and Ahmad, H. (2014). Graphene oxide-based waveguide polariser: From thin film to quasi-bulk, *Optics Express*, **22**, pp. 11090–11098.
8. Feth, J. R., and Chang, C. L. (1986). Metal-clad fiber-optic cutoff polarizer, *Optics Letters*, **11**, pp. 386–388.
9. Andreev, A. Ts., Kozlov, V. A., Kuznetsov, A. V., and Maksimov, A. V. (1993). Single-mode fiber polarizers for the spectral range 0.6–1.6 μm, *Quantum Electronics*, **7**, pp. 617–619.
10. Feng, J., Zhao, Y., Lin, X. W., Hu, W., Xu, F., and Lu, Y. Q. (2011). A transflective nano-wire grid polarizer based fiber-optic sensor, *Sensors*, **11**, pp. 2488–2495.
11. Si, K. J., Sikdar, D., Chen, Y., Eftekhari, F., Xu, Z., Tang, Y., Xiong, W., Guo, P., Zhang, S., Lu, Y., Bao, Q., Zhu, W., Premaratne, M., and Cheng, W. (2014). Giant plasmene nanosheets, nanoribbons, and origami, *ACS nano*, **8**, pp. 11086–11093.
12. Hosaka, T., Okamoto, K., and Edahiro, T. (1983). Fabrication of single-mode fiber-type polarizer, *Optics Letters*, **8**, pp. 124–126.
13. Mikhailov, S., and Ziegler, K. (2007). New electromagnetic mode in graphene, *Physical Review Letters*, **99**, pp. 016803.
14. Bludov, Y. V., Vasilevskiy, M. I., and Peres, N. M. R. (2012). Tunable graphene-based polarizer, *Journal of Applied Physics*, **112**, pp. 084320.
15. Bao, Q., and Loh, K. P. (2012). Graphene photonics, plasmonics, and broadband optoelectronic devices, *ACS Nano*, **6**, pp. 3677–3694.
16. Kim, J. T., and Choi, S.-Y. (2011). Graphene-based plasmonic waveguides, *Optics Express*, **19**, pp. 24557–24562.

17. Kim, J. T., and Choi, C.-G. (2012). Graphene-based polymer waveguide polarizer, *Optics Express*, **20**, pp. 3556–3562.
18. Pei, C., Yang, L., Wang, G., Wang, Y., Jiang, X., Hao, Y., Li, Y., and Yang, J. (2015). Broadband graphene/glass hybrid waveguide polarizer, *Photonics Technology Letters, IEEE*, **27**, pp. 927–930.
19. Hsu, K.-Y., Ma, S.-P., Chen, K.-F., Tseng, S.-M., and Chen, J.-I. (1997). Surface-polariton fiber polarizer: Design and experiment, *Japanese Journal of Applied Physics*, **36**, pp. L488.

Chapter 5

Graphene-Based Photodetectors

Shaojuan Li,[a] Yan Liu,[a] Hui Ying Hoh,[b] and Qiaoliang Bao[a,b]

[a]*Institute of Functional Nano & Soft Materials (FUNSOM), Soochow University, 199 Ren-Ai Road, Suzhou Industrial Park, Suzhou Jiangsu 215123, China*
[b]*Department of Materials Science and Engineering, Monash University, 22 Alliance Lane, Building 69, Clayton, 3800, Victoria, Australia*
qiaoliang.bao@gmail.com

5.1 Introduction

Ultrafast photodetectors have many potential applications, ranging from environment monitoring to telecommunications, sensing, and basic research. To date, commercialized photodetectors are dominated by IV and III-V semiconductors, e.g., Si, Ge, GaAs, and GaN. However, the intrinsic bandgap of standard semiconductors restricts the photoresponsivity of the detectors to specific light bandwidths. Novel materials with broadband absorption are, therefore, required for wideband, tunable detection. Unlike conventional semiconductor materials, graphene, a 2D carbon nanosheet, shows attractive mechanical and photoelectrical properties, particularly wavelength-independent light absorption, outstanding high carrier mobility as well as ease of fabrication process [6]. Therefore, graphene has been

Graphene Photonics, Optoelectronics, and Plasmonics
Edited by Qiaoliang Bao, Hui Ying Hoh, and Yupeng Zhang

ISBN 978-981-4745-60-4 (Hardcover), 978-1-315-19667-1 (eBook)
www.panstanford.com

the foremost 2D atomic crystal investigated for photodetectors that show substantial advances in broadband detection, fast response, high cutoff frequency, and strain tolerance. The first reported graphene photodetector, which is a phototransistor, was reported in 2009. T. Mueller and F. Xia studied the photoelectric response of graphene field effect transistor. For a common metal–graphene–metal structure, carrier transfer from the metal contacts to graphene enables a bandgap bending and creates a built-in internal electric field at the interface. Driven by the electric field, photo-exited electron–hole pairs in graphene will be separated and then extracted under bias voltages, generating an electric output [22]. Using near-field scanning imaging, they showed that the strongest photocurrent spots occur where the metal contacts with graphene due to the existence of local electric field near the interface. The photocurrent gradually decreases as the distance increases to hundreds of nanometers away from the contacts, as shown in Fig. 5.1a. Later in the same year, they realized ultrafast photodetection using graphene phototransistor. The bandwidth of the device reached 40 GHz, and the maximum operation frequency is predicted to be as high as 500 GHz in theory [31]. This pioneer work is the first step toward exploring the full potential of graphene in broadband and ultrafast photodetection. In 2010, by using asymmetric electrodes in an inter-digitated metal–graphene–metal photodetector, the mirror symmetry of built-in electric field in the vicinity of two metal contacts was broken, enhancing the separation of the photocarriers; thus, performances of the graphene photodetector were improved with a responsivity of 6.1 mAW^{-1} in 10 $Gbit{\cdot}s^{-1}$ optical data link [23], as shown in Figs. 5.1b,c, which approved that graphene photodetector is a promising candidate for applications in high-speed optical communication.

It is worth noting that the lack of a bandgap of intrinsic graphene and limited light absorption in one atomic layer ($\pi\alpha$ = 2.3%) have greatly limited the responsivity of pure graphene-based photodetectors to a few mAW^{-1}, which is much lower than that of commercial Si photodiode [23, 31, 36]. In order to overcome these limitations, great efforts have been paid to improve the main figures of merits of graphene-based detectors, i.e., the photoconductive gain

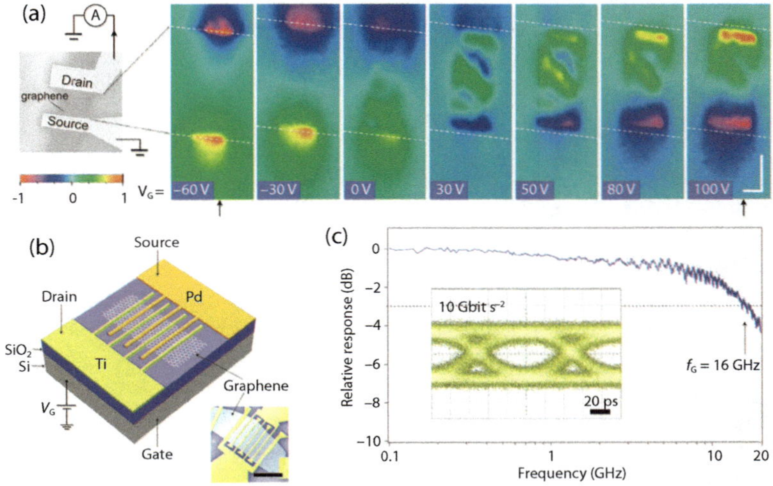

Figure 5.1 (a) Scanning photocurrent mapping image of graphene field effect transistor. Seven photocurrent images taken at gate biases between −60 and +100 V are shown on the right. The dashed lines indicate the edges of the source and drain electrodes. The two scale bars on the bottom of the very right image are both 1 μm long. Reprinted with permission from Ref. 22, Copyright 2009 by the American Physical Society. (b) Schematic diagram of asymmetrical inter-digital photodetector. (c) Broadband photoresponse measurement of the device in (b). Figures (b) and (c) reprinted with permission from Ref. 23, Copyright 2010, Nature Publishing Group.

and responsivity [2, 3, 5, 13, 17, 19, 21, 27, 33, 34, 35]. To date, the rapid development of graphene-based photodetection has mainly focused on two aspects. The first aspect is the enhancement of the light absorption in graphene. Another aspect is to strengthen the electron–hole separation and extraction. In this chapter, various approaches will be addressed separately, including plasma-enhanced graphene photodetectors, resonant cavity–enhanced graphene photodetectors, waveguide-integrated graphene photodetectors, and van der Waals heterostructure photodetectors. We will focus on the working principles and performance of the device. In the end of this chapter, conclusions and outlook in this field will be addressed.

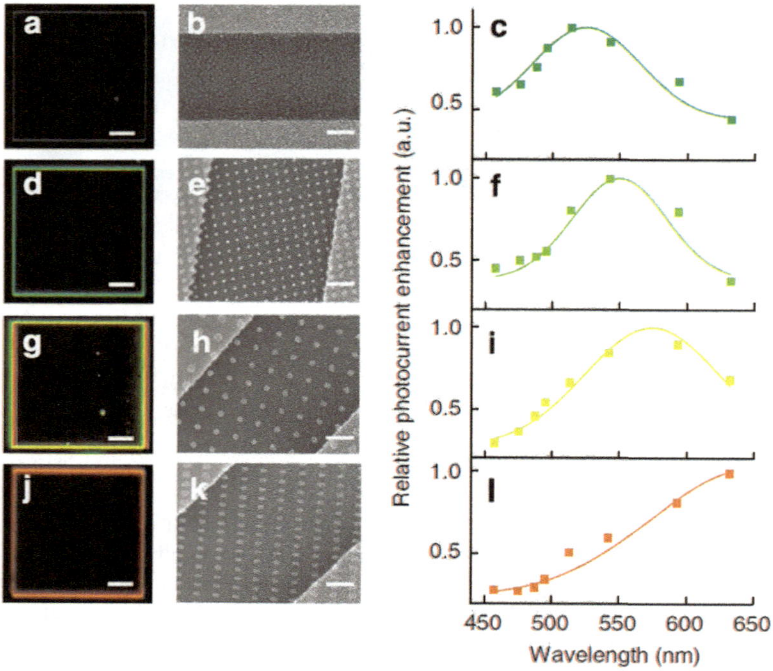

Figure 5.2 Multicolor photodetection using graphene devices coupled with different plasmonic nanostructures. (a) Dark-field image of a 50 × 50 μm^2 area of 18 nm diameter nanoparticle array obtained by thermally annealing 4 nm Au thin film. (b) Scanning electron microscopy (SEM) image of a representative device with the nanoparticle array, and (c) the corresponding spectral response of the photocurrent enhancement. (d) Dark-field image of a 50 × 50 μm^2 area of 50 nm diameter and 30 nm height nanodisk array. (e) SEM image of a representative device with the 50 nm diameter nanodisk array, and (f) the corresponding spectral response of the photocurrent enhancement. (g) Dark-field image of a 50 × 50 μm^2 area of 100 nm diameter and 30 nm height nanodisk array. (h) SEM image of a representative device with the 100 nm diameter nanodisk array, and (i) the corresponding spectral response of the photocurrent enhancement. (j) Dark-field image of a 50 × 50 μm^2 area of 100 nm length, 50 nm width, and 30 nm height nanorod array. (k) SEM image of a representative device with the nanorod array, and (l) the corresponding spectral response of the photocurrent enhancement. Scale bars are 10 μm in (a), (d), (g), and (j), and 400 nm in (b), (e), (h), and (k). The plasmonic nanostructures in (d), (g), and (j) are obtained by using e-beam lithography to define Au patterns. Reprinted by permission from Macmillan Publishers Ltd: *Nature Communications*, Ref. 19, Copyright 2011.

5.2 Plasma-Enhanced Graphene Photodetectors

Because of the inherently low absorption of normal incident light by graphene (2.3% for monolayer graphene), early studies on graphene-based photodetection focused on the enhancement of light absorption. Plasmonic coupling has been demonstrated to be a highly efficient solution [2, 19]. It has been shown that by placing noble metal nanostructures on the surface of a graphene sheet, these nanostructures can transform incident optical energy into plasmon resonance due to the local plasmonic enhancement effect, leading to a localized electrical field, which plays an important role in the generation, separation, as well as transport of photo-exited carriers, thus rendering improved device performance in terms of photoresponsivity. Most interestingly, selective photoresponse at different wavelengths can be achieved by means of changing the shapes, sizes, and thicknesses (i.e., resonance frequency) of the metallic plasmonic nanostructures. In 2011, Duan et al. utilized this strategy to fabricate highly sensitive multicolor photodetectors by coupling a high-density monolayer of gold nanoparticles with graphene, as shown in Fig. 5.2 [19]. An ultrahigh external quantum efficiency up to 1.5% was obtained (at zero source drain and gate biases), which is about one order of magnitude higher than previously reported graphene photodetectors (~0.1–0.2% at zero bias) [23, 30].

In addition to metallic plamonic nanostructures, plasmonic clusters made from polymer are also capable of effectively enhancing the direct excitation of electron–hole pair in monolayer graphene [4]. Figure 5.3 shows a plasmonic antenna–graphene photodetector consisting of fano-resonant plasmonic clusters sandwiched between two monolayers of graphene in a planar device. The photodetector achieved up to 20% internal quantum efficiency in the visible and near-infrared regions of the spectrum. The enhanced photoelectric response can be mainly attributed to two mechanisms. The first mechanism is that the hot carriers produced in the plasmonic nanostructures can be transferred into the conduction band of graphene through plasma relaxation, resulting in the increase in carrier concentration in graphene sheet. Another mechanism is that the excitation of electron—hole pairs in graphene is directly caused by the intense near field, thereby increasing the photocurrent.

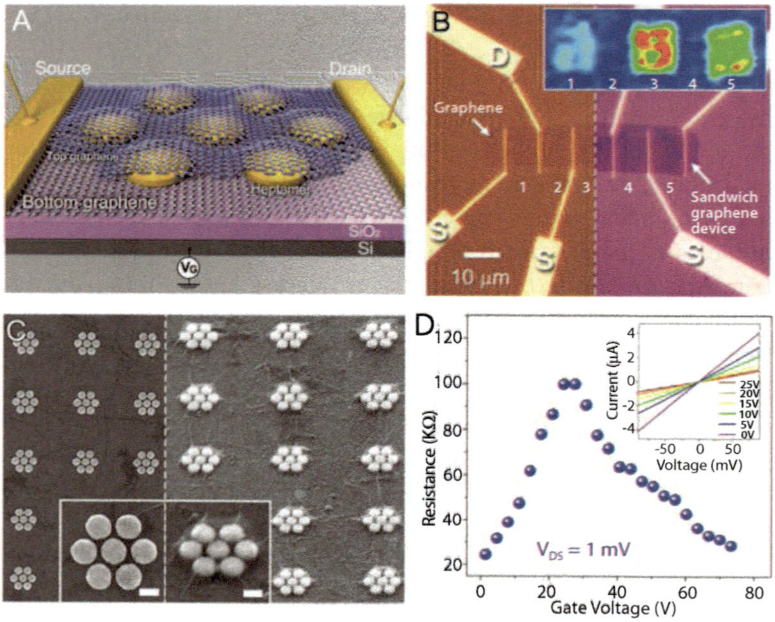

Figure 5.3 Plasmonic antenna–graphene photodetector consisting of Fano-resonant plasmonic clusters sandwiched between two monolayers of graphene in a planar device. (a) Schematic illustration of a single gold heptamer sandwiched between two monolayer graphene sheets. V_G is the gate voltage used to electrostatically dope the graphene. (b) Optical microscopy image of the as-fabricated device before (left) and after (right) deposition of the second graphene layer. Inset: Raman mapping (G mode) of device areas 1–5 under 785 nm excitation. (c) SEM image of heptamer array fabricated in area 3, as indicated in (b). (d) Electrical transport characteristic at a drain bias of 1 mV. Inset: I–V plots for various gate voltages V_G from 0 to 25 V. Reprinted with permission from Ref. 4, Copyright 2012, American Chemical Society.

5.3 Resonant Cavity–Enhanced Graphene Photodetectors

Among variant approaches to improve the light absorption of graphene photodetectors, embedding graphene into an optical cavity is ideally suited for obtaining highly efficient optical absorption in graphene, benefitting from the large increase in the optical field

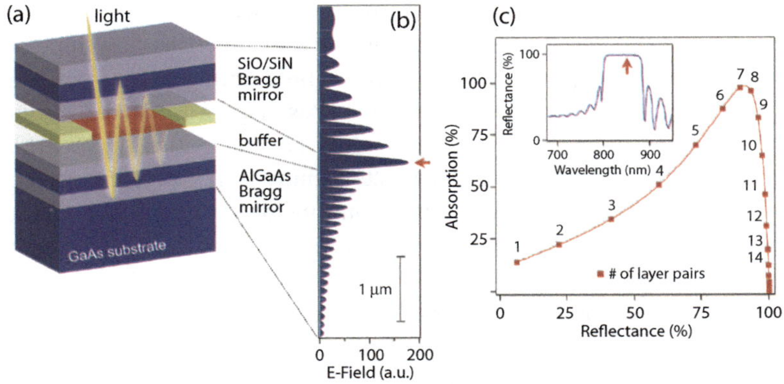

Figure 5.4 Schematic drawing of a graphene microcavity photodetector. Distributed Bragg mirrors form a high-finesse optical cavity. The incident light is trapped in the cavity and passes multiple times through the graphene. The graphene sheet is shown in red, and the metal contacts are in yellow. (b) Electric field amplitude inside the cavity. (c) Spectral response of the single-layer graphene device. The dashed lines show calculation results: reflection R (red), transmission T (green), and absorption A (blue). The solid lines are measurement results: reflection (red) and photocurrent (blue). A strong and spectrally narrow photoresponse is observed at the cavity resonance (855 nm wavelength). Inset: Theoretical result for normal incidence light. Reprinted with permission from Ref. 7, Copyright 2012, the ACS AuthorChoice.

in the middle of the resonant cavity [3, 5, 7]. The resonant cavity–induced confinement enables powerful controlling of the optical absorption of graphene at the designed wavelength controlled by the spectral properties of the microcavity. In 2012, the first demonstration of monolithic integration of a graphene transistor with a planar, optical microcavity was reported by Engel et al. [3]. In comparison with the non-confined graphene transistor, a 20-fold enhancement of photocurrent is demonstrated for the microcavity-controlled graphene detector. Moreover, it was found that the optical microcavity can effectively modify the electron transport characteristics of the internal graphene transistor, thus controlling the photocurrent generation of the device.

The Fabry–Perot microcavity, consisting of two distributed Bragg mirrors, is an ideal choice for microcavity-confined optoelectronic devices because the reflectivity in the cavity can be very well controlled and can reach values near unity. By inserting graphene at the position where the maximum optical field occurs in between the two Bragg mirrors, optical absorption was largely enhanced in the graphene sheet because incident photons can be confined inside the Fabry–Perot microcavity and reflected many times, prolonging the interaction time between graphene and incident phonons. In 2012, Mueller et al. demonstrated a monolithically integrating graphene photodetector with a Fabry–Perot microcavity [7]. The optical absorption reached as high as 60%, that means more than 60% of the incident light can be absorbed, which is a 26-fold enhancement compared to the non-microcavity graphene photodetector. The fabricated devices, therefore, also have a relatively high responsivity (21 mAW^{-1}). Figures 5.4a–c show the diagram of a Fabry–Perot microcavity–integrated graphene photodetector, the distribution of electric field intensity in the cavity, and spectral response of the device, respectively.

5.4 Waveguide-Integrated Graphene Photodetector

Though the above approaches can greatly improve the responsivity of graphene-based photodetectors, they restrict photodetection to narrow bands. More recent effort has been paid to develop other routes to produce high-efficiency graphene absorption without compromising broad spectral bandwidth and high speed. Since the interaction between graphene and normal incident light is relatively low, instead of utilizing normal incident light, coupling graphene with an in-plane evanescent field in an optical waveguide would be an alternative choice that allows the increase in the interaction length through coupling between graphene and the evanescent field in the optical waveguide beneath graphene. So far, significant achievements have been made in graphene–silicon optical waveguide-integrated optoelectronic devices [8, 10–12, 17, 20, 29]. In 2013, three research groups reported their latest process on graphene-based photodetectors and similar device structures were

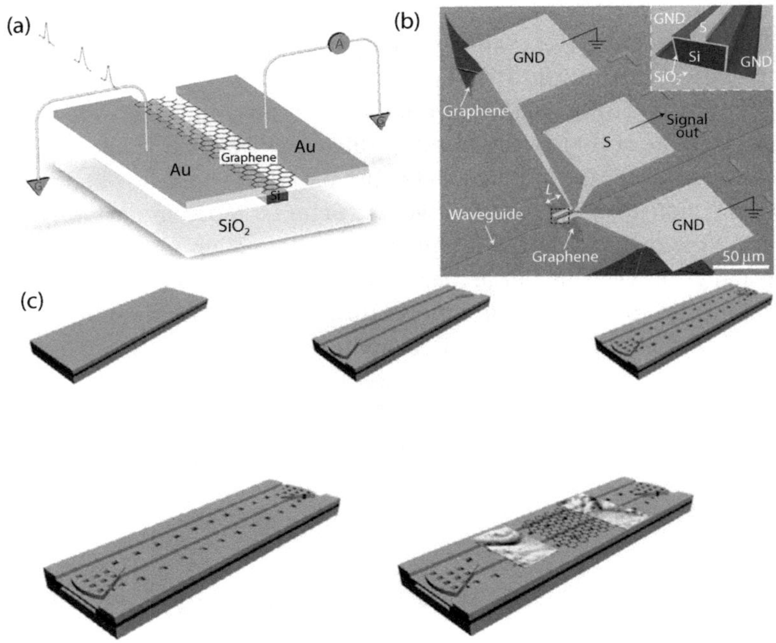

Figure 5.5 (a, b) Schematic diagrams of graphene–silicon-integrated waveguide photodetector. (c) Fabrication process Up, from left to right: SOI substrate, waveguide defined by first step EBL and DRIE process, via holes and focusing sub-wavelength grating defined by second step EBL and DRIE process Bottom left and right: suspended membrane waveguide after wet etching to remove silica, photodetector after transferring graphene and electrode fabrication. Figures (a), (b), and (c) reprinted by permission from Macmillan Publishers Ltd: *Nature Photonics*, Refs. 26, 29, and 8, respectively, Copyright 2013.

implemented by integrating graphene on the top of silicon optical waveguide [8, 26, 29], as shown in Fig. 5.5. The unique figures of this kind of structures lie in that the waveguide restricts and transmits the incident light in a silicon waveguide on a silicon-on-insulator (SOI) substrate, enabling absorption of evanescent light that propagates parallel to the graphene sheet over a long distance and excitation of photocarriers and hot carriers in the graphene sheet.

Graphene photodetectors composed of a graphene/waveguide configuration show great advantages especially in the high responsivity in near-infrared and mid-infrared operational range from 1.2 μm to 8 μm. For example, high photoresponsivity of 0.1 AW^{-1} between 1450 nm and 1590 nm [8] and 0.13 AW^{-1} for 2750 nm light [29] has been reported. It is notable that these infrared detectors can operate at room temperature, which is still a great challenge for the state-of-the-art infrared detectors based on other low-bandgap semiconductors or quantum-well structures. More importantly, it is well known that metal–graphene–metal junction photodetectors usually suffer from large dark currents and thus operate with significant power consumption. However, the graphene/silicon waveguide-integrated photodetector exhibits a low dark current and shows high responsivity, benefitting from a naturally formed graphene/silicon heterojunction. These experimental studies pave the way for the realization of complementary metal-oxide semiconductor (CMOS)-compatible integration of electro-optical systems on a single chip.

5.5 Van der Waals Heterojunction Structure Photodetector

The research of graphene leads to the discovery of many other 2D materials beyond graphene, among which semiconducting transition metal dichalcogenides (TMDCs), e.g., MoS_2, WS_2, and WSe_2, have drawn a large amount of interest because they have a moderate bandgap (1~2 eV) compared to the zero bandgap of graphene. As mentioned previously, the gapless nature of intrinsic graphene has prevented the photodetectors from exhibiting efficient carrier separation or accumulation, which leads to poor device performance in terms of low photoresponsivity and low photoconductive gain [23, 31, 36]. An important breakthrough in the field of graphene-based photodetectors is the creation of heterostructures based on graphene and other 2D materials with a bandgap.

In addition to the existence of a bandgap, the alluring characteristics of TMDCs also include their transition from indirect bandgap semiconductor to direct bandgap semiconductor as their form turns from bulk into two dimensional. Furthermore, TMDCs are poised to be ideal photovoltaic materials owing to their excellent

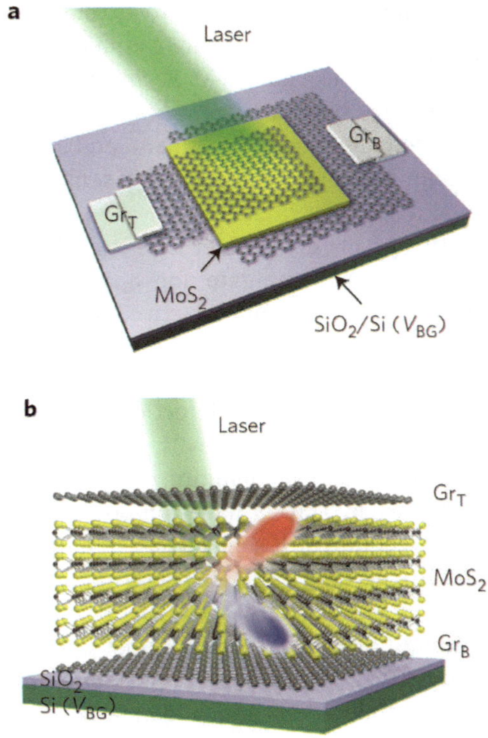

Figure 5.6 (a) Schematic illustration of the graphene–MoS_2–graphene vertical stacked device layout. (b) Schematic illustration of the side view of the device, with the semiconducting multilayer MoS_2 sandwiched between the Gr_T and Gr_B electrodes. Red and blue colors indicate electrons and holes, respectively. The silicon substrate can be used as a back-gate electrode with 300 nm SiO_2 as the dielectric layer. Reprinted by permission from Macmillan Publishers Ltd: *Nature Nanotechnology*, Ref. 34, Copyright 2013.

light absorption efficiency ($>10^7 m^{-1}$ across the visible range) and relatively high carrier mobility (10~200 $cm^2/V{\cdot}s$). In this respect, the performances of photodetectors based on graphene–MoS_2 or graphene–WS_2 heterojunction have achieved dramatic improvement in terms of high responsivity and sensitivity [27, 34, 35]. In such structures, graphene acts as transparent electrodes, and a TMDC layer is sandwiched between the two layers of graphene or between graphene and metal electrodes, forming a vertical transport barrier. The polarity of the photocurrent can be altered or switched on and off if the gate bias voltage is changed from negative to positive because of the change of Fermi energy in graphene. These phenomena can be quantitatively explained by the gate-tunable charge exchange between graphene and TMDC layers, and may lead to emerging graphene-based highly sensitive gate-tunable photodetectors and optical switch devices.

More interestingly, this novel heterostructure has been demonstrated to be a thriving candidate for applications in flexible photodetection, attributed to their attractive mechanical and photoelectrical properties. In 2013, the first demonstration was reported by Britnell et al. by utilizing graphene–WS_2–graphene stacking heterostructure for flexible photodetection [1], which presents extremely efficient photoresponse with high photoresponsivity above 0.1 AW^{-1} and an external quantum efficiency of above 30%.

As mentioned earlier, TMDCs have a moderate bandgap; thus, photocarriers excited in TMDCs have highly efficient generation and separation. To this point, planar heterojunction rather than vertical stacking heterostructure constructing from the TMDCs and graphene should also behave with remarkable optoelectronic functionality. In 2013, Roy et al. reported a planar, binary graphene/MoS_2 photodetector, which displays a responsivity of $5{\sim}10^8$ AW^{-1} at room temperature in the visible range of spectrum; that is the highest responsivity obtained in graphene-based photodetector then [27]. The study further proves that combining graphene with a thin layer of transition metal sulfides to form a heterostructure has great potential application for ultrasensitive photodetection.

5.6 Conclusion and Outlook

In conclusion, graphene-based photodetectors have been the hotspot of latest research due to their exceptional electrical and optical properties. To date, different types of graphene-based photodetectors with variant device configurations have been successfully demonstrated, which is sensitive to broadband spectrum ranging from UV to terahertz range, among which IR-sensitive graphene-based photodetectors are being extensively investigated owing to their unprecedented photoresponse even at room temperature.

For IR-sensitive graphene-based photodetectors, many approaches, as discussed in Sections 5.2–5.5, have been applied to improve performance and many impressive progresses have been achieved. Although the exact mechanism for photocurrent generation is still being debated [15, 25, 32], a junction is usually required to separate the photo-excited electron–hole pairs. Such junctions can be p-n junctions or heterojuctions. Whatever the junction is, a difference in the work function is critical while selecting a different material to combine with graphene. For example, creation of heterostructures based on graphene and other 2D materials offers the possibility to obtain unprecedented optoelectronic functionality. In the near future, the development of these structures may mainly suffer from the following problems: the precise synthesis of the junction and the high quality control.

The unique gapless band structure of graphene as well as its excellent transport properties makes graphene an appealing candidate for terahertz detection. Plasma-based graphene terahertz photodetectors with high performance at room temperature have also been realized [14, 16, 24]. For graphene-based terahertz photodetectors, interested readers can refer to Refs. 14, 24, and 32 of this chapter. In addition, graphene-based photodetectors on flexible substrates have also been developed recently [1, 9, 18, 28], implying that the material holds great potential for use in future large-area and flexible photodetectors.

References

1. Britnell, L., Ribeiro, R., Eckmann, A., Jalil, R., Belle, B., Mishchenko, A., Kim, Y.-J., Gorbachev, R., Georgiou, T., and Morozov, S. (2013). Strong light-matter interactions in heterostructures of atomically thin films, *Science*, **340**, pp. 1311–1314.
2. Echtermeyer, T., Britnell, L., Jasnos, P., Lombardo, A., Gorbachev, R., Grigorenko, A., Geim, A., Ferrari, A., and Novoselov, K. (2011). Strong plasmonic enhancement of photovoltage in graphene, *Nat Commun*, **2**, pp. 458.

3. Engel, M., Steiner, M., Lombardo, A., Ferrari, A. C., Löhneysen, H. V., Avouris, P., and Krupke, R. (2012). Light–matter interaction in a microcavity-controlled graphene transistor, *Nat Commun*, **3**, pp. 906.
4. Fang, Z., Liu, Z., Wang, Y., Ajayan, P. M., Nordlander, P., and Halas, N. J. (2012). Graphene-antenna sandwich photodetector, *Nano Lett*, **12**, pp. 3808–3813.
5. Ferreira, A., Peres, N., Ribeiro, R., and Stauber, T. (2012). Efficient graphene-based photodetector with two cavities, *Phys Rev B*, **85**, pp. 115438.
6. Fiori, G., Bonaccorso, F., Iannaccone, G., Palacios, T., Neumaier, D., Seabaugh, A., Banerjee, S. K., and Colombo, L. (2014). Electronics based on two-dimensional materials, *Nat Nanotechnol*, **9**, pp. 768–779.
7. Furchi, M., Urich, A., Pospischil, A., Lilley, G., Unterrainer, K., Detz, H., Klang, P., Andrews, A. M., Schrenk, W., and Strasser, G. (2012). Microcavity-integrated graphene photodetector, *Nano Lett*, **12**, pp. 2773–2777, (http://pubs.acs.org/doi/full/10.1021/nl204512x).
8. Gan, X., Shiue, R.-J., Gao, Y., Meric, I., Heinz, T. F., Shepard, K., Hone, J., Assefa, S., and Englund, D. (2013). Chip-integrated ultrafast graphene photodetector with high responsivity, *Nat Photon*, **7**, pp. 883–887.
9. Gao, Z., Jin, W., Zhou, Y., Dai, Y., Yu, B., Liu, C., Xu, W., Li, Y., Peng, H., Liu, Z., and Dai, L. (2013). Self-powered flexible and transparent photovoltaic detectors based on CDSE nanobelt/graphene Schottky junctions, *Nanoscale*, **5**, pp. 5576–5581.
10. Gosciniak, J., and Tan, D. T. H. (2013). Graphene-based waveguide integrated dielectric-loaded plasmonic electro-absorption modulators, *Nanotechnology*, **24**, pp. 1–9.
11. Kim, J. T., and Choi, C.-G. (2012). Graphene-based polymer waveguide polarizer, *Opt Express*, **20**, pp. 3556–3562.
12. Koester, S. J., and Li, M. (2012). High-speed waveguide-coupled graphene-on-graphene optical modulators, *Appl Phys Lett*, **100**, pp. 10.1063/1.4704663.
13. Konstantatos, G., Badioli, M., Gaudreau, L., Osmond, J., Bernechea, M., de Arquer, F. P. G., Gatti, F., and Koppens, F. H. (2012). Hybrid graphene-quantum dot phototransistors with ultrahigh gain, *Nat Nanotechnol*, **7**, pp. 363–368.
14. Koppens, F. H. L., Mueller, T., Avouris, P., Ferrari, A. C., Vitiello, M. S., and Polini, M. (2014). Photodetectors based on graphene, other two-dimensional materials and hybrid systems, *Nat Nanotechnol*, **9**, pp. 780–793.

15. Lemme, M. C., Koppens, F. H., Falk, A. L., Rudner, M. S., Park, H., Levitov, L. S., and Marcus, C. M. (2011). Gate-activated photoresponse in a graphene p–n junction, *Nano Lett*, **11**, pp. 4134–4137.

16. Li, J., Niu, L., Zheng, Z., and Yan, F. (2014). Photosensitive graphene transistors, *Adv Mater*, **26**, pp. 5239–5273.

17. Liu, M., Yin, X. B., Ulin-Avila, E., Geng, B. S., Zentgraf, T., Ju, L., Wang, F., and Zhang, X. (2011). A graphene-based broadband optical modulator, *Nature*, **474**, pp. 64–67.

18. Liu, N., Tian, H., Schwartz, G., Tok, J. B. H., Ren, T.-L., and Bao, Z. (2014). Large-area, transparent, and flexible infrared photodetector fabricated using p-n junctions formed by n-doping chemical vapor deposition grown graphene, *Nano Lett*, **14**, pp. 3702–3708.

19. Liu, Y., Cheng, R., Liao, L., Zhou, H., Bai, J., Liu, G., Liu, L., Huang, Y., and Duan, X. (2011). Plasmon resonance enhanced multicolour photodetection by graphene, *Nat Commun*, **2**, pp. 579.

20. Liu, Z. B., Feng, M., Jiang, W. S., Xin, W., Wang, P., Sheng, Q. W., Liu, Y. G., Wang, D. N., Zhou, W. Y., and Tian, J. G. (2013). Broadband all-optical modulation using a graphene-covered-microfiber, *Laser Phys Lett*, **10**, pp. 1–5.

21. Manga, K. K., Wang, S., Jaiswal, M., Bao, Q., and Loh, K. P. (2010). High-gain graphene-titanium oxide photoconductor made from inkjet printable ionic solution, *Adv Mater*, **22**, pp. 5265–5270.

22. Mueller, T., Xia, F., Freitag, M., Tsang, J., and Avouris, P. (2009). Role of contacts in graphene transistors: A scanning photocurrent study, *Phys Rev B*, **79**, pp. 245430.

23. Mueller, T., Xia, F., and Avouris, P. (2010). Graphene photodetectors for high-speed optical communications, *Nat Photon*, 4, pp. 297–301.

24. Otsuji, T., Tombet, S. A. B., Satou, A., Fukidome, H., Suemitsu, M., Sano, E., Popov, V., Ryzhii, M., and Ryzhii, V. (2012). Graphene-based devices in terahertz science and technology, *J Phys D Appl Phys*, **45**, pp. 515–517.

25. Park, J., Ahn, Y., and Ruiz-Vargas, C. (2009). Imaging of photocurrent generation and collection in single-layer graphene, *Nano Lett*, **9**, pp. 1742–1746.

26. Pospischil, A., Humer, M., Furchi, M. M., Bachmann, D., Guider, R., Fromherz, T., and Mueller, T. (2013). CMOS-compatible graphene photodetector covering all optical communication bands, *Nat Photon*, **7**, pp. 892–896.

27. Roy, K., Padmanabhan, M., Goswami, S., Sai, T. P., Ramalingam, G., Raghavan, S., and Ghosh, A. (2013). Graphene-MoS_2 hybrid structures

for multifunctional photoresponsive memory devices, *Nat Nanotechnol*, **8**, pp. 826–830.

28. Son, D. I., Yang, H. Y., Kim, T. W., and Park, W. I. (2015). Transparent and flexible ultraviolet photodetectors based on colloidal ZnO quantum dot/graphene nanocomposites formed on poly(ethylene terephthalate) substrates, *Compos Part B-Eng*, **69**, pp. 154–158.
29. Wang, X., Cheng, Z., Xu, K., Tsang, H. K., and Xu, J.-B. (2013). High-responsivity graphene/silicon-heterostructure waveguide photodetectors, *Nat Photon*, **7**, pp. 888–891.
30. Xia, F., Mueller, T., Golizadeh-Mojarad, R., Freitag, M., Lin, Y.-M., Tsang, J., Perebeinos, V., and Avouris, P. (2009). Photocurrent imaging and efficient photon detection in a graphene transistor, *Nano Lett*, **9**, pp. 1039–1044.
31. Xia, F., Mueller, T., Lin, Y.-M., Valdes-Garcia, A., and Avouris, P. (2009). Ultrafast graphene photodetector, *Nat Nanotechnol*, **4**, pp. 839–843.
32. Xia, F., Farmer, D. B., Lin, Y.-M., and Avouris, P. (2010). Graphene field-effect transistors with high on/off current ratio and large transport band gap at room temperature, *Nano Lett*, **10**, pp. 715–718.
33. Yan, H., Li, X., Chandra, B., Tulevski, G., Wu, Y., Freitag, M., Zhu, W., Avouris, P., and Xia, F. (2012). Tunable infrared plasmonic devices using graphene/insulator stacks, *Nat Nanotechnol*, **7**, pp. 330–334.
34. Yu, W. J., Liu, Y., Zhou, H., Yin, A., Li, Z., Huang, Y., and Duan, X. (2013). Highly efficient gate-tunable photocurrent generation in vertical heterostructures of layered materials, *Nat Nanotechnol*, **8**, pp. 952–958.
35. Zhang, W., Chuu, C.-P., Huang, J.-K., Chen, C.-H., Tsai, M.-L., Chang, Y.-H., Liang, C.-T., Chen, Y.-Z., Chueh, Y.-L., He, J.-H., Chou, M.-Y., and Li, L.-J. (2014). Ultrahigh-gain photodetectors based on atomically thin graphene-MoS_2 heterostructures, *Sci Rep-Uk*, **4**, doi:10.1038/srep03826.
36. Zhang, Y., Liu, T., Meng, B., Li, X., Liang, G., Hu, X., and Wang, Q. J. (2013). Broadband high photoresponse from pure monolayer graphene photodetector, *Nat Commun*, **4**, pp. 1811.

Chapter 6

Graphene for Solar Cells

Pengfei Li,[a] Yusheng Wang,[a] Qasim Khan,[a,b] Hui Ying Hoh,[c] and Qiaoliang Bao[a,c]

[a]*Institute of Functional Nano & Soft Materials (FUNSOM), Soochow University, 199 Ren-Ai Road, Suzhou Industrial Park, Suzhou Jiangsu 215123, China*
[b]*School of Electronic Science and Engineering, Southeast University, Nanjing 210096, Jiangsu, China*
[c]*Department of Materials Science and Engineering, Monash University, 22 Alliance Lane, Building 69, Clayton, 3800, Victoria, Australia*
qiaoliang.bao@gmail.com

6.1 Introduction

In recent years, solar energy has received extensive attention as a viable source of green energy. The variety of solar cells has increased considerably, from crystalline silicon solar cells, amorphous silicon thin-film solar cells to newer forms such as dye-sensitized solar cells (DSSCs) and organic solar cells. The specific electrical and optical properties are requisite to the material used in solar cells. In 2004, Geim et al. prepared graphene by the scotch tape method and sparked the unprecedented attention of graphene materials. Graphene, a single-atomic-layer material, has excellent intrinsic properties, such as large surface area (2630 m^2/g), outstanding carrier mobility of 20,000 $cm^2/(V \cdot s)$, large Young's modulus (1.0

Graphene Photonics, Optoelectronics, and Plasmonics
Edited by Qiaoliang Bao, Hui Ying Hoh, and Yupeng Zhang

ISBN 978-981-4745-60-4 (Hardcover), 978-1-315-19667-1 (eBook)
www.panstanford.com

TPa), excellent thermal conductivity of 5000 W/(m·k), and high light transmittance (97.7%) [1, 2]. These unique properties, along with exceptional electrical conductivity and mechanical flexibility, render graphene and its derivatives as potential candidates in the field of solar cells. To date, lots of impressive results have been reported, where graphene was used as electrodes, i.e., transparent cathodes as well as non-transparent anodes [3–5]. Graphene has also been used in DSSCs and organic photovoltaic cells (OPV), as light-harvesting material electron transport layer, hole transport layer (HTL), and interfacial layer in the tandem configuration [6–9]. In this chapter, we will review the application of graphene in solar cells.

6.2 Different Types of Graphene for Solar Cells

In this section, we briefly outline the types of graphene used in solar cells and the methods to produce them. Readers who are interested in a more detailed review on the various production methods may refer to Chapter 1. One of the most popular methods to obtain graphene is through chemical vapor deposition (CVD). Graphene sheets produced by CVD are usually used as transparent electrodes due to high transparency, lower concentration of defects, and ease of transfer to other substrates. At present, the material used for transparent electrodes is indium tin oxide (ITO), commonly known as conductive glass. Although conductive glass is widely used, it has some shortcomings, such as spontaneous diffusion of the metal ions, strong absorption in the infrared spectrum, and poor thermal stability. The use of ITO and fluorine tin oxide (FTO) is also increasingly problematic due to the shortage of indium, as well as limited transparency of ITO and FTO in the near-infrared region. In the subsequent sections, we will discuss encouraging results for graphene electrodes and we believe that in time, graphene will be a viable substitute for ITO or FTO.

Functionalized forms of graphene may also be produced from graphene oxide (GO) using chemical synthesis. The oxidation of graphite is usually performed through Hummer's method by reacting graphite with sulfuric acid, sodium nitrite, and potassium permanganate [10, 11]. One may adjust the type and concentration of functional groups on the graphene sheets obtained through subsequent reactions, whether through covalent bonding or non-covalent functionalization. The chemical and physical properties of the graphene sheets may, therefore, be easily tailored to the required applications [12]. Such forms of graphene are typically used in graphene-based nanocomposites for organic or hybrid solar cells. Till date, several reports have been demonstrated on the application of functionalized graphene for DSSCs, organic solar cells, and quantum dot solar cells (QDSCs).

6.3 Dye-Sensitized Solar Cells

6.3.1 Graphene as Transparent Electrodes

As early as 2008, Wang et al. used reduced graphene oxide (rGO) as the transparent conductive anode with the device structure of glass/rGO/TiO_2/dye/spiro-OMeTAD/Au (Fig. 6.1a) [13]. The corresponding energy level diagram of the DSSC device is shown in Fig. 6.1b. The graphene oxide was dip coated on quartz followed by thermal reduction. The conductivity of the graphene film is as high as 550 S/cm and with more than 70% transparency in the wavelength from 1000 to 3000 nm. This is the first demonstration of a solid state DSSC based on a graphene electrode, showing a short-circuit current density (J_{sc}) of 1.01 mA/cm^2 with an open-circuit voltage (V_{oc}) of 0.7 V, a fill factor (FF) of 0.36, and the power conversion efficiency (PCE) of 0.26%, which is lower than the FTO-based solid state DSSC (0.84%; Fig. 6.1c).

In 2010, Lei Jiang et al. introduced graphene as 2D bridges into the nanocrystalline electrodes of DSSCs, which brought a faster electron transport and a lower recombination, together with a higher light scattering [14]. On the basis of these advantages, the short-circuit current density was increased by 45% without sacrificing the open-circuit voltage, and the total conversion efficiency was 6.97%, which was increased by 39%, comparing with the nanocrystalline titanium dioxide photoanode. In 2013, Wang et al. used nitrogen-doped graphene (NDG) sheets [15]. It is found that chemical reduction and nitrogen doping of graphite oxide are achieved simultaneously via a simple hydrothermal process. Electrochemical

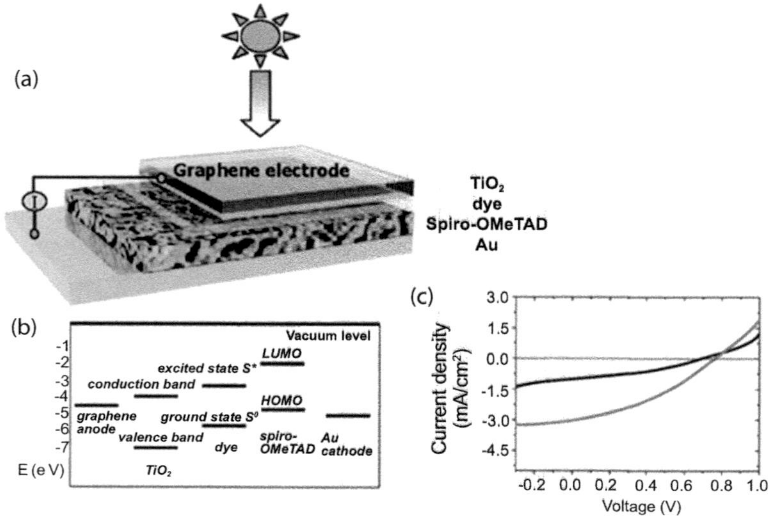

Figure 6.1 Illustration and performance of a DSSC based on graphene electrodes. (a) Schematic illustration of a DSSC using a graphene film as electrode. (b) The energy level diagram of the DSSC with the configuration of graphene/TiO_2 /dye/spiro-OMeTAD/Au. (c) Current density–voltage curves of the graphene-based cell (black) and FTO-based cell (red) illuminated under AM solar light (sun). Reprinted with permission from Ref. 13, Copyright 2008, American Chemical Society.

impedance spectroscopy analysis reveals that the charge-transfer resistance of NDG electrode decreases with increasing the NDG loading. The electrode with an NDG loading of 20 μgcm^2 shows a charge-transfer resistance of 0.9 Ωcm^2, which is much lower than that of pristine graphene electrode. The solar cell with NDG electrode shows an energy conversion efficiency of 7.01%, which is comparable to that of the cell with platinum electrode.

Recently, Jeon et al. prepared edge-selectively fluorinated graphene nanoplatelets (FGnPs) by a mechanochemically driven reaction between fluorine gas (20 vol% in argon) and activated carbon species from graphitic C–C bonds unzipped by high-speed stainless steel balls with a high kinetic energy [16]. Electrodes based on the resultant FGnPs demonstrate superb electrochemical performance with excellent stability/cycle life in DSSCs (FF: 71.5%;

J_{sc}: 14.44 mA/cm^2; V_{oc}: 970 mV; PCE: 10.01%). It is believed that there is still room for improvement of device performance using graphene as transparent electrodes. Readers may refer to the review by Chang et al. for more examples [17]. As researchers have fabricated graphene-based transparent electrodes with RSh <30 Ω/sq at $T = 90\%$, approaching the performance of conventional FTO and ITO electrodes, it is evident that graphene electrodes can enter the market. However, it is not clear if they will do so and if so when. The scalability of the technology and cost reductions must be demonstrated for graphene to be used in commercial devices.

With regards to DSSCs, extra constraints may arise depending on the electrolyte and solar cell design. For instance, metal hybrid electrodes would likely corrode in an iodine-containing electrolyte. In any case, using graphene materials in transparent electrodes is far from mature, and more advances will be made, particularly for applications in which the current TCFs do not work well, such as low-temperature substrates and flexible devices [18].

6.3.2 Graphene as Catalytic Counter Electrodes

Graphene has also been used for the catalytic counter electrode in DSSCs in order to replace the traditional expensive platinum (Pt) counter electrode. For example, Aksay et al. used functionalized graphene sheets (FGS) as a stand-alone catalyst in a DSSC to replace Pt [19]. The catalytic activity from FGS is attributed to its same level charge-transfer resistance with platinum at an applied bias, and the introduced oxygen-containing functional groups on FGS (Figs. 6.2a,b). The FGS-based cell achieves a high efficiency of 5%, which is only 10% lower compared to the corresponding Pt-based cell (Fig. 6.2c). Finally, they demonstrated that the FGS ink-casted plastic substrate can be used as an efficient counter electrode, eliminating the requirement for the common FTO substrate. Tailoring the functionalization or morphology of the FGS electrodes could decrease the charge-transfer resistance and also facilitate the low-cost production of the catalytic, flexible, and conductive counter electrodes for DSSCs. Grätzel et al. have also shown that graphene nanoplatelets (GNPs) demonstrated high electro-catalytic activity as a counter electrode material for the redox reaction of

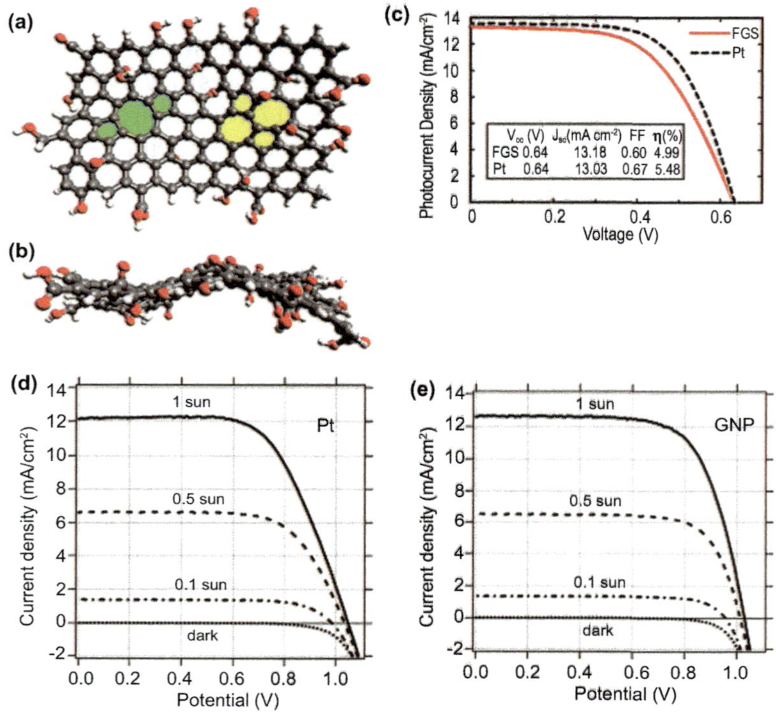

Figure 6.2 (a) Schematic illustration of functional groups and lattice defects on an FGS. Carbon, oxygen, and hydrogen atoms are gray, red, and white, respectively. (b) Side view emphasizing the topography of the FGS. (c) J–V characteristics of DSSCs using thermally decomposed chloroplatinic acid and FGS counter electrodes. Figures (a), (b), and (c) reprinted with permission from Ref. 19, Copyright 2010, American Chemical Society. (d) J–V characteristics of DSSCs with Pt counter electrode. (e) J–V characteristics of DSSCs with GNP counter electrode. Figures (d) and (e) reprinted with permission from Ref. 20, Copyright 2011, American Chemical Society.

$Co(L)_2$ (L = 6-(1H-pyrazol-1-yl)-2,2′-bipyridine) [20]. The exchange current density for the $Co^{2+/3+}(L)_2$ redox was observed to be one to two orders of magnitude larger than those for the conventional I^{3-}/I^- redox couple on the same electrode. In the $Co^{2+/3+}(L)_2$ redox system, the electro-catalytic activity of GNP-film cathode with optical transmission below 88% outperformed the activity of the Pt electrode. DSSCs with Y123 dye adsorbed on TiO_2 photoanode achieved the energy conversion efficiency of 8–10% for both GNP-

and Pt-based cathodes. Importantly, the cell with GNP cathode is superior to that with Pt cathode, particularly in the FF and the efficiency at higher illumination intensity (Figs. 6.2d,e).

In addition to pure graphene, graphene-based composites with controlled functionalization have also been developed for the catalytic counter electrodes. Chen et al. first realized the direct growth of vertically aligned carbon nanotubes (VACNTs) on $\approx 3\ \mu m$ thick graphene paper (GP) by a thermal CVD method [21]. The as-prepared freestanding VACNT/GP film possesses superb flexibility and durability, which is attributed to the highly flexible GP. When employed as the counter electrode in a DSSC, it displayed superior overall performance over the GP or the tangled CNT/GP film and only slightly lower efficiency compared to the Pt electrode. Therefore, it is highly likely that graphene can replace Pt in the near future.

6.4 Organic/Polymer Solar Cells

6.4.1 Functionalized Graphene as Electrode

Graphene films were fabricated by controlled pyrolysis from the botanical derivative camphor [22]. These films show very good transparency in a wide range of wavelengths (0.3–2 mm) in contrast to ITO. A P3HT:PCBM bulk heterojunction (BHJ) solar cell, which was fabricated on a graphene electrode with transparency of 81% at 550 nm and a conductivity of 357 S/cm, exhibited very good open-circuit voltage (V_{oc}), while the FF and short-circuit current density (J_{sc}) are affected with the high sheet resistance of the graphene film.

Polyethylene terephthalate (PET) flexible substrate was used for the transfer of chemically derived reduced GO films. The reduced GO films were used as the transparent, conductive electrode for solar cell devices with the RGO/PEDOT:PSS/P3HT:PCBM/TiO_2/Al structure [23]. Device performance is highly dependent on the charge transport efficiency of the reduced GO film, which in turn is determined by its sheet resistance. The highest PCE value obtained for the flexible RGO/PET-based solar cell devices was 0.78%. However, these devices showed excellent stability at high bending induced stress and were well maintained even at very high bending.

Graphene has also been successfully used as the transparent conductive anode for the flexible organic photovoltaic (OPV) cell with the structure of graphene/poly(3,4-ethylenedioxythiophene):poly(styrenesulfonate)(PEDOT:PSS)/copperphthalocyanine (CuPc)/C_{60}/bathocuproine(BCP)/Al (Fig. 6.3a) [24]. Graphene is first deposited onto the Si/SiO_2/Ni film by CVD and followed by transferring onto the transparent glass or polyethylene terephthalate (PET) substrates (Fig. 6.3b). The obtained graphene/PET film has a sheet resistance of 230 Ω/sq with a transparency of 72%. The solar cell based on CVD-grown graphene as the transparent conductive anode shows a PCE of 1.8%, which is comparable to the performance (1.27%) by a device made with the commonly used ITO electrode. Obviously, the flexibility of CVD-graphene device surpasses that of the ITO device because graphene can operate under bending up to 138° (Fig. 6.3c), whereas the latter degrade rapidly and irreversibly even under bending of 60° (Fig. 6.3d). Consistently, the decrease rate of the initial FF in the CVD-graphene device is much slower than that of the ITO one as the bending angle increased, and the FF of the ITO device rapidly decays to 0 when it is bent to around 60° (Fig. 6.3e). The lower performance due to the bending of the ITO device is caused by the generation of microcracks in the ITO film under mechanical stress, while no such cracks are observed on the CVD-graphene device (Fig. 6.3f). Readers are encouraged to read the review by Yin et al. for more examples [25]. These promising results suggest that graphene will see uses in the area of flexible OPVs in the near future.

6.4.2 Functionalized Graphene as Transport Layer

Graphene oxide was also used as the electron acceptor in solar cell device with an ITO/PEDOT:PSS/P3HT:GO/Al structure, which exhibited an increase in conductivity and J_{sc} but a decrease in V_{oc} [26]. A bilayer photovoltaic device based on the spin cast GP3HT/C_{60} hetero-structures showed a PCE of 0.61% under AM 1.5 illumination (100 mW/cm^2). Devices were prepared with P3HT covalently grafted GO sheets (GP3HT) via esterification between the CH_2OH-terminated P3HT and carboxylic groups in GO [7].

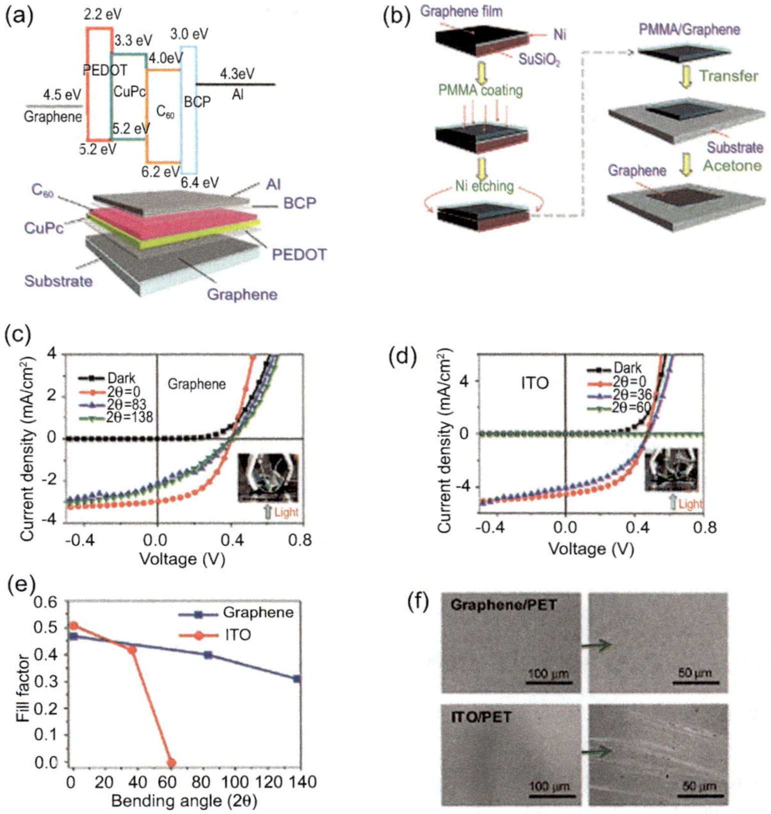

Figure 6.3 (a) Schematic representation of the energy level alignment (top) and the construction of heterojunction organic solar cell fabricated with graphene as anodic electrode: graphene/PEDOT/CuPc/C_{60} /BCP/Al. (b) The plots of current density versus voltage for (c) graphene and (d) ITO devices under 100 mW/cm^2 AM1.5G spectral illumination at different bending angles. Insets show the setup used in the experiments. (e) The plot of fill factor versus bending angle for the graphene and ITO devices. (f) Scanning electron microscopy (SEM) images of the surface structure of CVD-graphene (top) and ITO (bottom) devices after being subjected to the bending angles described in panels (c,d). Reprinted with permission from Ref. 24, Copyright 2010, American Chemical Society.

The thinnest GO thin films (~2 nm) deposited from neutral solutions are effective HTLs in solar cells [8]. The PCE values of devices with GO HTLs were comparable to devices fabricated with PEDOT:PSS, which is the most common anode interfacial layer to improve anode contact and to increase hole collection in solar cell devices. PEDOT:PSS is highly acidic in nature. In addition, its hygroscopic properties and inhomogeneous electrical properties render it a poor material for long-term stability compared to GO films. Energy band alignment obtained from work function and optical gap measurements of the GO thin films demonstrated that electron blocking and hole transport were facilitated. A semi-transparent solar cell that uses a top laminated graphene electrode on an inverted structure using a simultaneous thermal annealing/releasing technique was demonstrated [27]. This device exhibited a promising PCE of approximately 76% coupled with silver (Ag) electrode. The deposition of the graphene electrode was demonstrated to be fully compatible with the roll-to-roll production in future market integration for low-cost plastic solar cells.

6.4.3 Functionalized Graphene as Active Layer

To effectively tune the bandgap of graphene and facilitate their application in nanodevices, one approach is to convert the 2D graphene sheets into 0D graphene quantum dots (GQDs). Li et al. reported an electrochemical approach for the direct preparation of functional GQDs with a uniform size of 3–5 nm, which was stable in water for several months without any degradation [21]. Polymer solar cell with ITO/PEDOT:PSS/P3HT:GQDs/Al structure exhibited a PCE of 1.28%. GQDs could be environmentally friendly, cost effective, and more stable for photovoltaics (PVs) than the current organic materials, and blending with regioregular MEH-PPV or poly(3-hexylthiophene-2,5-diyl) polymer results in a significant improvement in the OPV characteristics as compared to graphene sheets (GSs) blended conjugated polymers [28]. Figure 6.4 (inset) shows the typical cell structure of GQDs-based solar cell, with ITO/PEDOT:PSS/P3HT: aniline(ANI)-GQDs/ LiF/Al structure, maximum values of $\eta = 1.14\%$, $V_{oc} = 0.61$ V, FF $= 0.53$, and $J_{sc} =$ 3.51 mA/cm^2 were obtained when the content of ANI-GQDs in P3HT

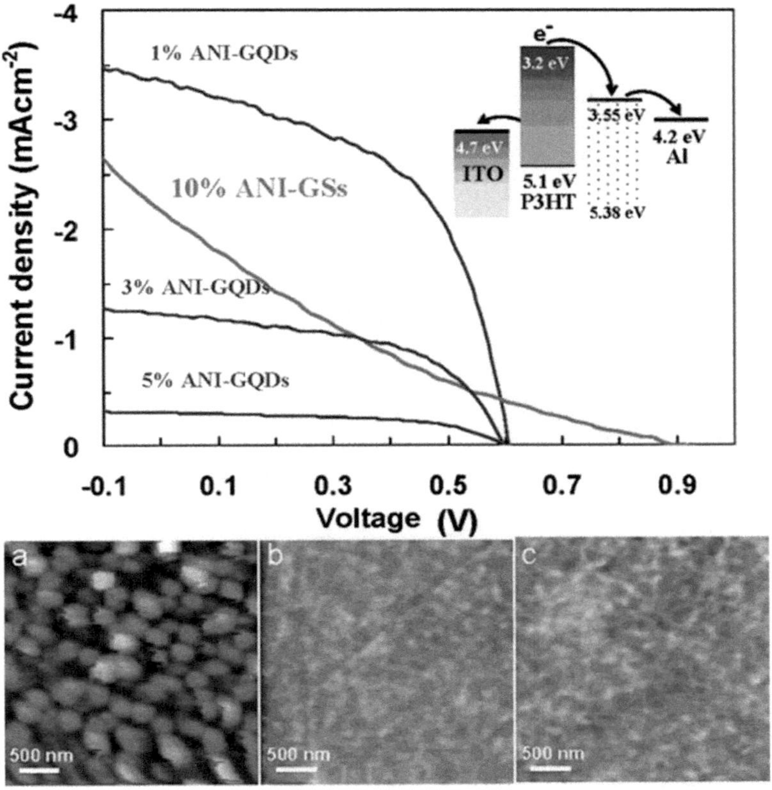

Figure 6.4 Above: J–V characteristics of the solar cells based on ANI-GQDs with different GQDs content and ANI-GS (optimized condition) annealed at 160° C for 10 min, in AM1.5G 100 mW illumination. Below: AFM images of (a) P3HT/ANI-GSs, (b) P3HT/ANI-GQDs, and (c) MEH-PPV/methylene blue-GQDs. Reprinted with permission from Ref. 28, Copyright 2011, American Chemical Society.

is 1 wt.%. The FF of the P3HT/1 wt.% ANI-GQDs hetrojunction device is much higher (0.53) as compared to ~0.33 for GSs, and 0.28 for 10 wt% ANI-GQDs.

As mentioned previously, GO can be readily tailored to various requirements owing to the presence of functional groups. For example, Chen et al. functionalized GO sheets with phenyl isocyanate, which changed the hydrophilic GO surface to a hydrophobic

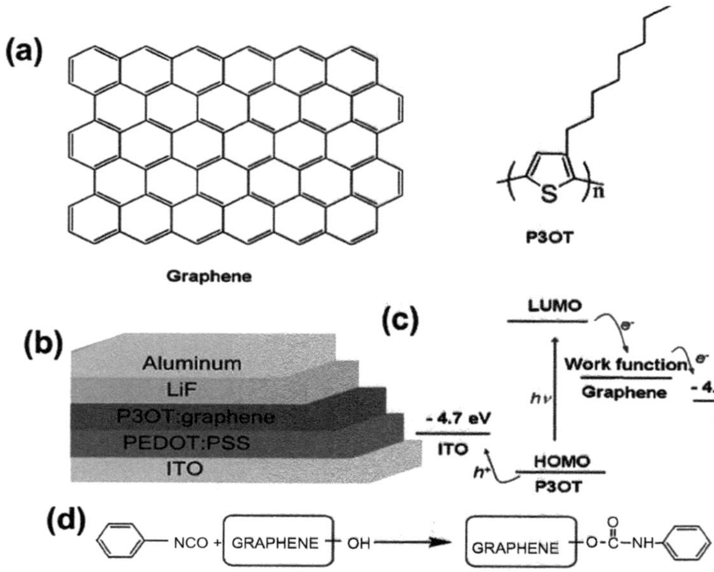

Figure 6.5 (a) The idealized chemical structures of graphene and P_3OT. (b) Schematic illustration of the device with P_3OT/graphene thin film as the active layer and the structure ITO/PEDOT:PSS (40 nm)/P_3OT:graphene (100 nm)/LiF (1 nm)/Al (70 nm). (c) Energy level diagram of P_3OT and SPFGraphene. (d) Schematic representation of the reaction of phenyl isocyanate with GO to form SPFGraphene. Reprinted from Ref. 6 with permission from John Wiley and Sons.

one [29]. The resultant solution-processed functionalized graphene (SPFGraphene) is mixed with poly(3-octylthiophene) (P_3OT) to form the P_3OT/SPFGraphene composites, which are then used as the active layer in the bulk heterojunction (BHJ) OPV device (Fig. 6.5). The annealing conditions are critical for the performance of the device, since annealing can remove the functional groups from graphene sheets and enhance the crystallinity of P_3OT. The optimized annealing condition (160°C for 20 min) is applied for the fabrication of an OPV device, which achieved the best PCE of 1.4%. This work indicated that the functional graphene can serve as a competitive alternative to [6,6]-phenyl C61-butyric acid methyl ester (PCBM) as the electron acceptor for high-performance OPV devices.

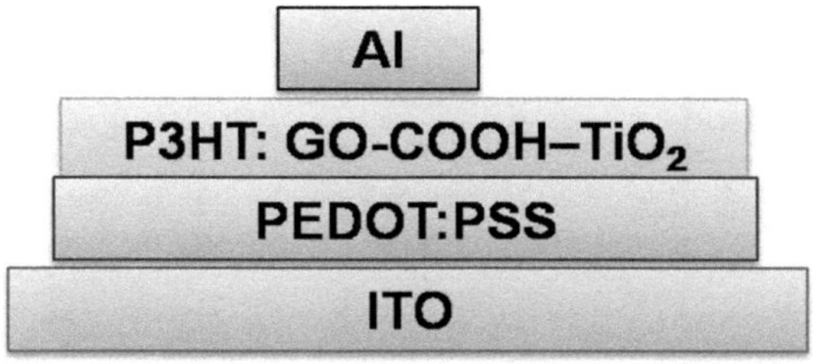

Figure 6.6 The schematic of the hybrid solar cell with ITO/PEDOT (50 nm)/P3HT: GO-COOH–TiO_2/Al structure.

Another challenge in the integration of graphene in solar cell devices is the incompatibility between graphene and PEDOT:PSS hole transporter, which subsequently causes device failure. As an alternative to PEDOT:PSS, the MoO_3 HTL was investigated to address the issue of surface immiscibility between graphene and the HTL [30]. MoO_3-modified graphene-based solar cell devices showed quite high J_{sc} of 11.6 mA/cm^2 and V_{oc} of 1 V could be obtained in parallel and series connection, respectively, in contrast to single photovoltaic cell efficiency, J_{sc} of 7.6 mA/cm^2 and V_{oc} of 0.58 V [30].

In another example, functionalized C_{60}-graphene hybrid as the electron acceptor in P3HT-based BHJ solar cells demonstrated a 2.5-fold increase in the PCE with an enhanced J_{sc} and V_{oc} voltage compared to those of the C_{60}/P3HT system under AM 1.5 illumination (100 mW/cm^2) [26]. The PCE value of C60-G:P3HT device (1.22%) was relatively low as compared to the P3HT:[60]PCBM (phenyl-C61- butyric acid methyl ester) cells (5–6%).

Furthermore, researchers have successfully developed the carboxylic acid graphene oxide and TiO_2 nanorod composites (GO-COOH–TiO_2) by a simple ligand exchange process. On the surface of a carboxylic acid–functionalized graphene oxide (GO-COOH) sheet, TiO_2 nanorods were uniformly anchored followed by the oleic acid ligands insulation cleaning from the surface of TiO_2 nanorods. The composite is a good acceptor material in hybrid solar cells and can effectively quench the photoluminescence of P3HT. The P3HT/GO-COOH–TiO_2 hybrid solar cell (Fig. 6.6) showed marked improvement compared to a P3HT/oleic acid capped TiO_2 solar cell [31, 32].

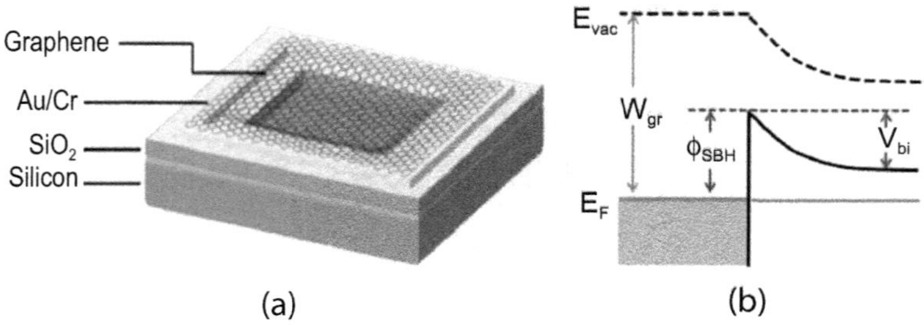

Figure 6.7 Graphene/n-type semiconductor Schottky junction solar cell: (a) schematic of the device structure, (b) corresponding energy diagram. Reprinted with permission from Ref. 2, Copyright 2012, American Chemical Society.

6.5 Graphene-Silicon Hybrid Solar Cells

Due to the near-zero bandgap and high conductivity of graphene, the graphene/n-type semiconductor heterojunction can be taken as a metal/semiconductor Schottky junction, assuming that the work function difference between the graphene and the semiconductor is sufficiently large. The mechanism of such Schottky junction solar cell can be understood qualitatively by plotting the energy band diagram.

Figure 6.7 shows the energy diagram of a graphene/n-type semiconductor Schottky junction solar cell under illumination and the schematic of the corresponding device structure. Due to the work function difference between the graphene (FG) and semiconductor (FS), a built-in potential forms in the semiconductor near the interface. Under light illumination above the bandgap, the photo-generated holes (h^+) and electrons (e) are separated and driven toward the Schottky electrode (graphene film) and semiconductor layer, respectively, by the built-in electric field. When the solar cell is under open-circuit conditions, the separated photo-generated electrons and holes will produce an open-circuit voltage V_{OC}. When the solar cell is short-circuited, the extracted photo-generated carriers will transit through the external circuit, generating a short-circuited current I_{SC}. Compared to its traditional counterpart, in the graphene-based Schottky junction solar cell, more light can go through the Schottky electrode and excite electron–hole pairs in the semiconductor, which will help to increase the optical to electrical energy conversion efficiency.

The solar energy conversion efficiencies are 1.65% and 1.34% for devices with junction areas 0.1 cm^2 and 0.5 cm^2, respectively [33]. Although the efficiency of such GS/n-Si Schottky junction devices is still far lower than that of pure silicon thin-film solar cells ($\approx$12%), it provides an improved understanding of the effects on solar cell performance from the electronic coupling, surface passivation, doping, and junction formation. Besides, more advanced and efficient graphene/silicon-based architectures with improved solar cell performance are possible [33]. For example, Miao et al. reported that the performance of Schottky solar cells based on the graphene–Si system can be greatly enhanced after doping of graphene [2]. The graphene sheet was first doped with bis(trifluoromethanesulfonyl)amide [$(CF_3SO_2)_2NH$] (TFSA). This TFSA-doped (p-type) graphene exhibited lower sheet resistance. Importantly, the TFSA doping raised the work function of graphene, thus increasing the built-in potential between the doped graphene and n-Si in solar cells. As shown in the J–V characteristics, J_{SC}, V_{OC}, and FF increase from 14.2 to 25.3 mA/cm^2, 0.43 to 0.54 V, and 0.32 to 0.63, respectively, after doping of the graphene sheets with TFSA. Increases in J_{SC} and V_{OC} boost the PCE from 1.9% to 8.6% (Fig. 6.8), which is the highest PCE reported for graphene-based solar cells to date. The pristine device shows an energy conversion efficiency (EQE) near 50% in the wavelength range of 400–850 nm, indicating the significant electron–hole pair generation and the subsequent facile collection of electrons and holes by the corresponding electrodes. After the TFSA doping, the EQE was significantly increased to a value of $\approx$65% in the aforementioned wavelength range, which is due to the more efficient charge separation and charge collection as a result of the increased built-in potential and reduced sheet resistance.

Based on the aforementioned discussion, graphene, which works as the electron acceptor after being functionalized with organic materials or as the Schottky junction metal layer combined with the inorganic semiconductors, has presented promising performance in solar harvesting. Since the performance is still low compared to

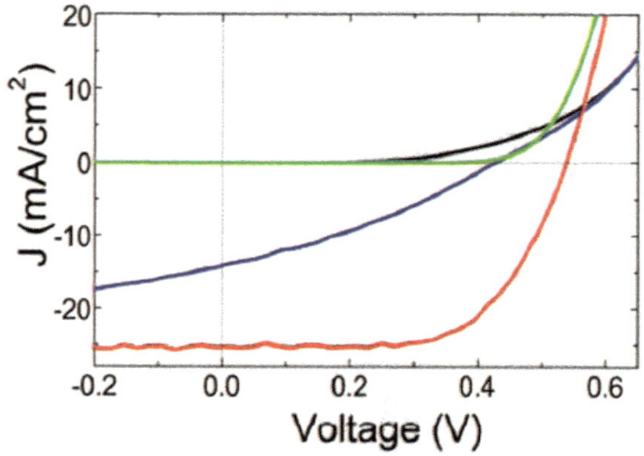

Figure 6.8 J–V characteristics before (black and blue) and after (green and red) doping of the graphene sheets with TFSA. Reprinted with permission from Ref. 2, Copyright 2012, American Chemical Society.

the current equivalent solar cells, a lot of work should be done for the enhancement of performance, such as through the design of graphene-based architectures with specific physical properties at heterojunctions, the chemistry at interface of the graphene-based hybrids, and the exploration of advanced materials and processing technologies [34, 35]. Combining low cost, mass production, high electron mobility with efficient charge transport, and 2D features with facile functionality, graphene is expected to play an increasingly important role in both research and industry for sustainable clean energy by partially or even fully replacing silicon and other organic photovoltaic materials in photovoltaic cells in the future.

6.6 Conclusion

This chapter summarizes the recent progress in the research on graphene-based solar cells. Considering that graphene can be manufactured by different methods with varying properties, it is a highly versatile material for applications in energy conversion and other optoelectronic devices. Device performance may be optimized through tailored functionalities on graphene sheets or sophisticated device structures. Continued research in this field should be very fruitful.

References

1. Zhan, D., Yan, J., Lai, L., Ni, Z., Liu, L., and Shen, Z. (2012). Engineering the electronic structure of graphene, *Advanced Materials*, **24**, pp. 4055–4069.
2. Miao, X., Tongay, S., Petterson, M. K., Berke, K., Rinzler, A. G., Appleton, B. R., and Hebard, A. F. (2012). High efficiency graphene solar cells by chemical doping, *Nano Letters*, **12**, pp. 2745–2750.
3. Bae, S., Kim, H., Lee, Y., Xu, X., Park, J.-S., Zheng, Y., Balakrishnan, J., Lei, T., Kim, H. R., and Song, Y. I. (2010). Roll-to-roll production of 30-inch graphene films for transparent electrodes, *Nature Nanotechnology*, **5**, pp. 574–578.
4. Lin, T., Huang, F., Liang, J., and Wang, Y. (2011). A facile preparation route for boron-doped graphene, and its CdTe solar cell application, *Energy and Environmental Science*, **4**, pp. 862–865.
5. Yin, Z., Wu, S., Zhou, X., Huang, X., Zhang, Q., Boey, F., and Zhang, H. (2010). Electrochemical deposition of ZnO nanorods on transparent reduced graphene oxide electrodes for hybrid solar cells, *Small*, **6**, pp. 307–312.
6. Liu, Z., Liu, Q., Huang, Y., Ma, Y., Yin, S., Zhang, X., Sun, W., and Chen, Y. (2008). Organic photovoltaic devices based on a novel acceptor material: Graphene, *Advanced Materials*, **20**, pp. 3924–3930.
7. Liu, Z., He, D., Wang, Y., Wu, H., and Wang, J. (2010). Graphene doping of P3HT:PCBM photovoltaic devices, *Synthetic Metals*, **160**, pp. 1036–1039.
8. Li, S.-S., Tu, K.-H., Lin, C.-C., Chen, C.-W., and Chhowalla, M. (2010). Solution-processable graphene oxide as an efficient hole transport layer in polymer solar cells, *ACS Nano*, **4**, pp. 3169–3174.
9. Tong, S. W., Wang, Y., Zheng, Y., Ng, M. F., and Loh, K. P. (2011). Graphene intermediate layer in tandem organic photovoltaic cells, *Advanced Functional Materials*, **21**, pp. 4430–4435.
10. Wu, J. B., Becerril, H. A., Bao, Z. N., Liu, Z. F., Chen, Y. S., and Peumans, P. (2008). Organic solar cells with solution-processed graphene transparent electrodes, *Applied Physics Letters*, **92**, pp. 263302.

11. Xu, Q. J., Song, T., Cui, W., Liu, Y. Q., Xu, W. D., Lee, S. T., and Sun, B. Q. (2015). Solution-processed highly conductive PEDOT:PSS/AgNW/GO transparent film for efficient organic-Si hybrid solar cells, *ACS Applied Materials and Interfaces*, **7**, pp. 3272–3279.

12. Loh, K. P., Bao, Q. L., Ang, P. K., and Yang, J. X. (2010). The chemistry of graphene, *Journal of Materials Chemistry*, **20**, pp. 2277–2289.

13. Wang, X., Zhi, L., and Müllen, K. (2008). Transparent, conductive graphene electrodes for dye-sensitized solar cells, *Nano Letters*, **8**, pp. 323–327.

14. Yang, N., Jin, Z., Dan, W., Chen, Y., and Lei, J. (2010). Two-dimensional graphene bridges enhanced photoinduced charge transport in dye-sensitized solar cells, *ACS Nano*, **4**, pp. 887–894.

15. Wang, A., Long, L., Zhao, W., Song, Y., Humphrey, M. G., Cifuentes, M. P., Wu, X., Fu, Y., Zhang, D., Li, X., and Zhang, C. (2013). Increased optical nonlinearities of graphene nanohybrids covalently functionalized by axially-coordinated porphyrins, *Carbon*, **53**, pp. 327–338.

16. Jeon, I. Y., Ju, M. J., Xu, J., Choi, H. J., Seo, J. M., Kim, M. J., Choi, I. T., Kim, H. M., Kim, J. C., Lee, J. J., Liu, H. K., Kim, H. K., Dou, S., Dai, L., and Baek, J. B. (2015). Edge-fluorinated graphene nanoplatelets as high performance electrodes for dye-sensitized solar cells and lithium ion batteries, *Advanced Functional Materials*, **25**, pp. 1170–1179.

17. Chang, D. W., Choi, H. J., Filer, A., and Baek, J. B. (2014). Graphene in photovoltaic applications: Organic photovoltaic cells (OPVs) and dye-sensitized solar cells (DSSCs), *J. Mater. Chem. A*, **2**, pp. 12136–12149.

18. Roy-Mayhew, J. D., and Aksay, I. A. (2014). Graphene materials and their use in dye-sensitized solar cells, *Chemical Reviews*, **114**, pp. 6323–6348.

19. Roy-Mayhew, J. D., Bozym, D. J., Punckt, C., and Aksay, I. A. (2010). Functionalized graphene as a catalytic counter electrode in dye-sensitized solar cells, *ACS Nano*, **4**, pp. 6203–6211.

20. Kavan, L., Yum, J. H., and Grätzel, M. (2011). Graphene nanoplatelets out-performing platinum as the electrocatalyst in co-bipyridine-mediated dye-sensitized solar cells, *Nano Letters*, **11**, pp. 5501–5506.

21. Li, Y., Hu, Y., Zhao, Y., Shi, G., Deng, L., Hou, Y., and Qu, L. (2011). An electrochemical avenue to green-luminescent graphene quantum dots as potential electron-acceptors for photovoltaics, *Advanced Materials*, **23**, pp. 776.

22. Kalita, G., Matsushima, M., Uchida, H., Wakita, K., and Umeno, M. (2010). Graphene constructed carbon thin films as transparent electrodes for

solar cell applications, *Journal of Materials Chemistry*, **20**, pp. 9713–9717.

23. Choi, Y.-Y., Kang, S. J., Kim, H.-K., Choi, W. M., and Na, S.-I. (2012). Multilayer graphene films as transparent electrodes for organic photovoltaic devices, *Solar Energy Materials and Solar Cells*, **96**, pp. 281–285.
24. Gomez De Arco, L., Zhang, Y., Schlenker, C. W., Ryu, K., Thompson, M. E., and Zhou, C. (2010). Continuous, highly flexible, and transparent graphene films by chemical vapor deposition for organic photovoltaics, *ACS Nano*, **4**, pp. 2865–2873.
25. Yin, Z., Zhu, J., He, Q., Cao, X., Tan, C., Chen, H., Yan, Q., and Zhang, H. (2014). Graphene-based materials for solar cell applications, *Advanced Energy Materials*, **4**, pp. 1–19.
26. Hill, C. M., Zhu, Y., and Pan, S. (2011). Fluorescence and electroluminescence quenching evidence of interfacial charge transfer in poly (3-hexylthiophene): Graphene oxide bulk heterojunction photovoltaic devices, *ACS Nano*, **5**, pp. 942–951.
27. Lee, Y.-Y., Tu, K.-H., Yu, C.-C., Li, S.-S., Hwang, J.-Y., Lin, C.-C., Chen, K.-H., Chen, L.-C., Chen, H.-L., and Chen, C.-W. (2011). Top laminated graphene electrode in a semitransparent polymer solar cell by simultaneous thermal annealing/releasing method, *ACS Nano*, **5**, pp. 6564–6570.
28. Gupta, V., Chaudhary, N., Srivastava, R., Sharma, G. D., Bhardwaj, R., and Chand, S. (2011). Luminescent graphene quantum dots for organic photovoltaic devices, *Journal of the American Chemical Society*, **133**, pp. 9960–9963.
29. Chen, T., Hu, W., Song, J., Guai, G. H., and Li, C. M. (2012). Interface functionalization of photoelectrodes with graphene for high performance dye-sensitized solar cells, *Advanced Functional Materials*, **22**, pp. 5245–5250.
30. Park, H., Brown, P. R., Buloyic, V., and Kong, J. (2012). Graphene as transparent conducting electrodes in organic photovoltaics: Studies in graphene morphology, hole transporting layers, and counter electrodes, *Nano Letters*, **12**, pp. 133–140.
31. Liu, J. C., Wang, Y. J., and Sun, D. D. (2013). Enhanced performance of hybrid solar cells using carboxylic acid-functionalized graphene oxide supported TiO_2 nanorod composites, *Materials Letters*, **95**, pp. 178–181.
32. Kaniyoor, A., and Ramaprabhu, S. (2012). An optically transparent cathode for dye sensitized solar cells based on cationically functionalized and metal decorated graphene, *Nano Energy*, **1**, pp. 757–763.

33. Li, X., Zhu, H., Wang, K., Cao, A., Wei, J., Li, C., Jia, Y., Li, Z., Li, X., and Wu, D. (2010). Graphene-on-silicon Schottky junction solar cells, *Advanced Materials*, **22**, pp. 2743–2748.

34. Jiao, K., Duan, C., Wu, X., Chen, J., Wang, Y., and Chen, Y. (2015). The role of MoS_2 as an interfacial layer in graphene/silicon solar cells, *Physical Chemistry Chemical Physics*, **17**, pp. 8182–8186.

35. Song, Y., Li, X., Mackin, C., Zhang, X., Fang, W., Palacios, T., Zhu, H., and Kong, J. (2015). Role of interfacial oxide in high-efficiency graphene–silicon Schottky barrier solar cells, *Nano Letters*, **15**, pp. 2104–2110.

Chapter 7

Graphene Plasmonics

Qingyang Xu,[a] Yao Lu,[a] Jian Yuan,[a] Yupeng Zhang,[b] and Qiaoliang Bao[a, b]

[a]*Institute of Functional Nano & Soft Materials (FUNSOM), Soochow University, 199 Ren-Ai Road, Suzhou Industrial Park, Suzhou Jiangsu 215123, China*
[b]*Department of Materials Science and Engineering, Monash University, 22 Alliance Lane, Building 69, Clayton, 3800, Victoria, Australia*
qiaoliang.bao@gmail.com

7.1 Introduction

Plasmons—the collective oscillations of valence electrons in conducting materials—possess a number of appealing properties for photonic technologies [1], the most salient of which are (1) their small spatial extension compared with the light wavelength, which has been exploited to achieve improved imaging resolution [2]; (2) their strong interaction with light, which is evidenced by a centenary tradition of generating colors through plasmon-supporting metal nanoparticle suspensions [3]; and (3) the huge optical enhancements produced by this strong interaction, which upon external illumination result in near-field intensities $>10^5$ times larger than the incident light intensity, as inferred from

Graphene Photonics, Optoelectronics, and Plasmonics
Edited by Qiaoliang Bao, Hui Ying Hoh, and Yupeng Zhang

ISBN 978-981-4745-60-4 (Hardcover), 978-1-315-19667-1 (eBook)
www.panstanford.com

surface-enhanced Raman scattering (SERS) measurements [4]. Control over the spectral and spatial properties of these collective excitations has advanced at an impressive pace in recent years [5, 6]. Equally impressive are their applications to ultrasensitive detection down to the single-molecule level [7], improved photovoltaics [8], nanoscale photometry [9], cancer therapy [10], and nonlinear optics [11], among other feats.

Unlike the metal plasmonic materials, the one-atom-thick graphene is so thin ($\approx$0.34 nm) that the semi-infinite interface model cannot be used to describe the plasmonic properties. At the beginning of the 19th century, it was considered that graphene was unstable and cannot exist due to strong thermal fluctuation of 2D materials. However, the rise of graphene has come since it was successfully obtained by the mechanical exfoliation of graphite and deposited on a Si wafer capped with 300 nm thickness SiO_2. Subsequently, the plasmonics based on graphene has become one of the most exciting research topics.

7.2 Theoretical Investigations

Plasmon excitations are intrinsic charge or current oscillations coupled via Coulomb interaction. One of them called surface plasmon polariton (SPP) is a plasmon bound to an interface with a dielectric medium. It originates from plasma physics, which is a very mature field and was studied extensively for well over a century. The dual identity of these polaritons—half optical wave, half electronic oscillation—offers the possibility of simultaneously carrying out optical imaging and probing the electronic properties of the surface of a material.

The phenomenon about plasmon was first reported by Wood [12] in 1902: he observed uneven distribution of light in a diffraction spectrum. Nearly 40 years later, Fano [13] theoretically revealed that Wood's phenomenon, which relied on the subsequently excited Sommerfeld-type EM waves with large tangential momentum on a metallic surface, cannot be described by Rayleigh's approximation [14]. In 1879, Crookes [15] reported the fourth fundamental state of matter in which positive ions and negative electrons or ions coexist, and he called it "radiant matter." After that, Langmuir studied the oscillations in this ionized state and named it plasma [16]. Soon afterward, he and Tonks [17] found that plasmas can sustain ion and electron oscillations and form a dilatational wave of the electron density. This wave can be quantized as plasma oscillations, which can be called plasmon. Subsequently, Pines and Bohm did a lot experimental and theoretical work on the origin of characteristic energy losses of fast electrons when they pass through foils and suggested that some of the energy losses should be due to the excitation of plasmon, which was a collective behavior [18]. They found the resonant frequency of plasmon in bulk plasma as:

$$\omega_0 = \left(\frac{ne^2}{m\varepsilon_0}\right)^{\frac{1}{2}} \tag{7.1}$$

where n and m are the electron density and mass, respectively, and ε_0 is the permittivity of vacuum. In 1957, Ritchie [19] found that an anomalous energy loss happened both at and below the oscillation frequency of plasmon when an electron traveled along thin film, and he suggested that this phenomenon should depend on the interface of the materials. Following these previous work, Stern and Ferrell studied the plasma vibration of the degenerate electron gas related to the material surface and first named them surface plasmons (SPs) in 1960 [20]. SPs are the collective oscillations of charges at the surface of plasmonic materials. Although the plasmons inside the materials decay quickly owing to the heavy energy loss, the SP can propagate quite a long distance.

With the rapid expansion of plamonic field, researchers found that SPs can be excited or coupled with different energies, i.e., photons, electrons, and phonons [21–25]. For example, SPs can couple with photons and form the composite particles of SPPs. Theoretically, the dispersion relationship between the frequency and wave vector for SPPs propagating along the interface of semi-infinite medium and dielectric can be obtained by surface mode solutions of Maxwell's equations under appropriate boundary conditions [26], and the non-radiation solution is the dispersion of SPPs:

$$k_{\mathrm{sp}} = k_0\sqrt{\frac{\varepsilon_{\mathrm{m}}\varepsilon_{\mathrm{d}}}{\varepsilon_{\mathrm{m}} + \varepsilon_{\mathrm{d}}}} \tag{7.2}$$

where ε_m and ε_d are the relative permittivities of medium and dielectric, and k_0 is the wave vector of light in free space. It should be noted that SPPs cannot be directly excited by light in an ideal semi-infinite plasmonic medium; the wave vectors of SPs and light should be matched to excite SPPs. Based on the dispersion relationship, k_{sp} can be complex with the positive real part standing for propagation and negative one standing for decay, which always relies on ε_m. Considering metallic materials, ε_m can be derived from the Drude model:

$$\varepsilon_m = 1 - \frac{\omega_0^2}{\omega^2 + i\tau^{-1}\omega} \tag{7.3}$$

where τ represents the relaxation time of the electrons in metal [27]. Thanks to plasmonics, many bottlenecks are broken such as nanophotonics [28], metamaterials [29], photovoltaic devices [30], and sensors [31].

Plasmons can be excited in many conventional metals such as Au, Ag, Cu, and Al, and these metals were regarded as the best plasmonic materials in the past for a long time. However, these noble metals suffer strong energy losses, and it is not easy to control the SP in metals. These shortcomings limit the development and application of plasmonics. Since graphene was exfoliated successfully by Novoselov and Geim in 2004 [32], it brings a revolution of material. The 2D carbon sheet has already been shown to possess many unique properties. Graphene has extremely high quantum efficiency for light–matter interactions, is strongly optically nonlinear, and contains plasmons with unusual properties [33]. The electrons in graphene behave like massless Dirac fermions, which results in extraordinary properties, e.g., carriers with high mobility and long mean free path, gate-tunable carrier densities, anomalous quantum Hall effects, and so on [34]. Owing to the special nature, the intrinsic graphene plasmons have an important advantage compared with plasmons in noble metals as they can be tuned by gating, doping, and chemical means. Moreover, the low losses and the efficient wave localization up to mid-infrared frequencies also lead it to be a promising alternative in future applications [35].

7.2.1 Electron Behavior in Two Dimensions

To investigate graphene plasmons, we first introduce the electron behavior in 2D electron systems (2DESs) and 2D plasmons. In 2DES-like graphene, electrons do not move as independent particles; their motions are highly correlated as a result of pairwise interactions. These can be written as a potential: $u\left(r_{ij}\right) = u\left(\left|\vec{r}_i - \vec{r}_j\right|\right)$, where $\vec{r}_{ij} = \vec{r}_i - \vec{r}_j$ represents the relative distance. For graphene with two sides exposed to different mediums (dielectric constants are written as ε_1 and ε_2, respectively), the potential $u\left(r_{ij}\right) = e^2/\varepsilon r_{ij}$, where $\varepsilon = (\varepsilon_1 + \varepsilon_2)/2$. The electron gas in graphene can be described at low energies by the following continuum-model Hamiltonian:

$$\hat{H} = v_{\mathrm{F}} \sum_i \sigma \cdot P_i + \frac{1}{2} \sum_{i \neq j} \frac{e^2}{\varepsilon \left|r_i - r_j\right|} \tag{7.4}$$

where $v_{\mathrm{F}} \approx 10^6\ \mathrm{ms}^{-1}$ represents the Fermi velocity, $P_i = -i\hbar\nabla_{r_i}$ is the canonical momentum of the electron, $\sigma = \left(\sigma_x \cdot \sigma_y\right)$ is a 2D vector of the Pauli matrices. This formula holds for electrons with given spin and valley indices. So the relative importance of electron–electron interactions can be quantified by the ratio between the second term's magnitude and the first term's magnitude. The distance between electrons in a doped graphene is about the inverse of the Femi wave number, that is, k_{F}^{-1}, the second term is about $e^2 k_{\mathrm{F}}/\varepsilon$ and the kinetic energy is of the order of $\hbar v_{\mathrm{F}} k_{\mathrm{F}}$. So the ratio between the two quantities defines a dimensionless parameter, which is usually called the graphene fine-structure constant:

$$\alpha_{ee} = \frac{e^2}{\varepsilon \hbar v_{\mathrm{F}}} \tag{7.5}$$

The graphene fine-structure constant can be tuned by changing the dielectric environment surrounding graphene. From these, we can conclude that the electrons in doped graphene can interact quite strongly with each other, and this effect has a great influence on graphene plasmon [33].

7.2.2 Dispersion Relation of Graphene Plasmons

The dispersion relation of plasmons is very important for graphene plasmonics, and numerous achievements have been made in theory and experiment, such as semi-classical model and random-phase approximation (RPA). The plasmon dispersion in 2DESs can be understood in the long-wavelength limit ($\ll k_F$) on the basis of hydrodynamics [33]. In this limit, the electron dynamics can be described macroscopically by defining two collective variables [25]: the deviation of the electron density from its average value, $(\mathbf{r}, t)$, and the associated current density, $\delta\mathbf{j}(\mathbf{r}, t)$. For $|\delta n/n \ll 1|$, the plasmons are described by the following linearized Euler equation of motion:

$$\frac{\partial \mathbf{j}(\mathbf{r},t)}{\partial t} = -\frac{D}{\pi e^2}\nabla_2 \int d^2\mathbf{r}' \frac{e^2}{\varepsilon\,|\mathbf{r}-\mathbf{r}'|}\delta n(\mathbf{r}',t) \tag{7.6}$$

Combined with the continuity equation, this leads to an equation for the Fourier component of $\delta n(\mathbf{r}, t)$:

$$\left[\omega^2 - \frac{D}{\pi e^2} q^2 u_{\mathrm{q}}\right] \delta n\,(\mathbf{q},\omega) = 0 \tag{7.7}$$

where D is the Drude weight. This equation implies that the plasmon frequency $\omega_{\text{plasmon}}\,(q) = \sqrt{2Dq/\varepsilon}$. From this equation, we can find a special nature of plasmon oscillation in two dimensions: $\omega_{\text{plasmon}}\,(q) \propto \sqrt{q}$. Considering the ordinary Schrödinger fermions with mass m_{b}, and the Drude weight $D = \pi e^2 n/m_{\mathrm{b}}$, we can get the well-known result $\omega_{\text{plasmon}} = \sqrt{2\pi n e^2 q/(\varepsilon m_{\mathrm{b}})}$. This result does not fit well for graphene. For the massless Dirac fermions (MDFs) in graphene, the Drude weight is given by $D_{\text{MDF}} = 4E_{\mathrm{F}}\sigma_{\text{uni}}/\hbar$, and the plasmon frequency in doped graphene in the long-wavelength limit can be written as:

$$\omega_{\text{plasmon}}\,(q) = \sqrt{\frac{8E_{\mathrm{F}}\sigma_{\text{uni}}q}{\hbar\varepsilon}} \tag{7.8}$$

So the Dirac plasmon frequency scales like $E_{\mathrm{F}}^{1/2} \propto n^{1/4}$ and contains Plank's constant. When the plasmon energy becomes larger than the threshold for interband transitions, plasmons become strongly Landau damped. Therefore, the compression of the SP wavelength relative to the excitation wavelength is decided by the fine-structure constant:

$$\lambda_{\text{plasmon}}/\lambda_0 \approx \frac{2\alpha E_{\mathrm{F}}}{\epsilon\hbar\omega} \approx \alpha \tag{7.9}$$

When $q \approx k_F$, the hydrodynamic approach is not adapted to describe the plasmon dispersion relation. Actually, the key ingredient of a fully quantum mechanical calculation of plasmon modes is the retarded density–density response function $\chi_m(q, \omega)$ [36]. This function is given by $\chi_m^{RPA}(q, \omega) = \chi_m^0(q, \omega)/\varepsilon(q, \omega)$ in the framework of linear response theory and within the RPA, where $\chi_m^{(0)}(q, \omega)$ is the non-interacting response function and $\varepsilon(q, \omega)$ is the RPA dynamical dielectric function.

Surface plasmons are the collective oscillation of charges, which are silent in plasmonic materials unless drawn by a definite amount of energy and momentum, such as photons, electrons, and phonons. Within these interactions, many special phenomena were found gradually in different plasmonic materials and flourishing the promising field of plasmonics. As far as we know, photons [36] and electrons [37] can couple with SPs in the form of quasi-particles, which are very interesting in optoelectronic information technique and condensed matter physics.

7.2.3 Surface Plasmon Polaritons

Although the existence of SPs in graphene and the interaction between SPs and low-energy electrons had already been revealed by electron energy-loss spectroscopy (EELS) and other spectroscopic studies, the direct visualization of propagating and localized graphene SPs is still highly desirable. SPs can be excited by photons and form SPPs, and it is easy to be probed. The excitation of SPs in graphene also faces the problem of mismatch of energy and momentum with those of light in free space compared with metals.

Thus, prism, topological defects, and periodic corrugations [38] were adopted to solve this problem. Recently, Fei et al. [39] observed SPPs in gated graphene on SiO_2 substrate with the scanning near-field optical microscope. To realize the match of wave vectors, they illuminated the sharp tip of an atomic force microscope (AFM) with a focused infrared beam, as shown in Fig. 7.1. At the fixed frequency of incident IR light, the SPPs are excited by the illuminated tip and can propagate along the graphene sheet; then the waves are reflected, interfered, and damped at the graphene edges. The propagated SPPs can be interfered by the reflected ones and form standing waves so that the wavelength of SPPs can be conveniently measured from the

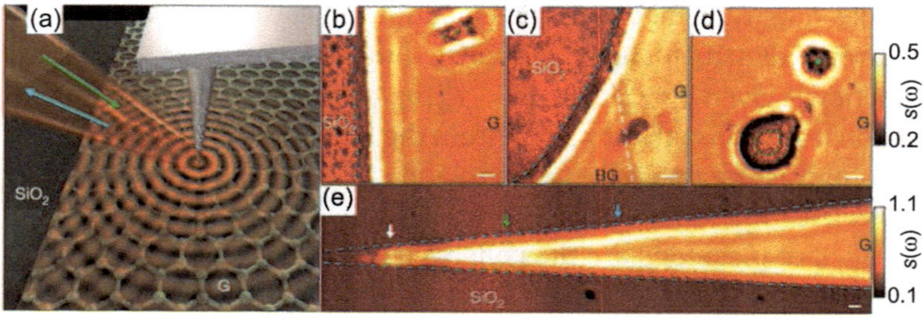

Figure 7.1 Infrared nano-imaging results of SPs in graphene. (a) Diagram of an infrared nano-imaging experiment at the surface of graphene. The green and blue arrows display the incident and back-scattered light, respectively. Concentric red circles illustrate plasmon waves induced by the illuminated tip of AFM. (b–e) Images of infrared amplitude and the interference pattern close to graphene edges (blue dashed lines) and defects (green dashed lines and green dot), and at the boundary between single (G) and bilayer (BG) graphene (white dashed line). Scale bars, 100 nm. Reprinted with permission from Ref. 39, Copyright 2012, Nature Publishing Group.

standing waves. Moreover, both the amplitude and wavelength of SPPs in graphene can be altered by varying the gate voltage.

7.2.4 Surface Plasmons Coupled with Phonons

In EELS experiments, acoustic-like quasi-linear dispersions of plasmons are found in the long-wavelength limit, which is a strange quantum behavior in graphene. As early as in 1959, Kohn had predicted that $\partial\omega(q)/\partial q|\, q = 2k_{\mathrm{F}} = \infty$ for the phonon wave mode in ordinary 2D metals due to the strong electron–phonon interactions, which was called Kohn anomaly [40]. By Raman spectroscopy, Kohn anomalies are also found in graphene with a breakdown of the Born–Oppenheimer approximation. Therefore, the phonons interacting with electron might be responsible for the quasi-linear dispersion. Because of the chirality in graphene, Tse et al. found four distinct Kohn anomalies, which are different from the metals [41]. After that, researchers achieved the breakthrough from graphene on polar substrates. Within the angle-resolved EELS, a strong plasmon–phonon coupling was found in epitaxial graphene on SiC(0001) [42]. As interpreted, surface state charges on the SiC are transferred into empty π^* states in the graphene sheet, and surface optical

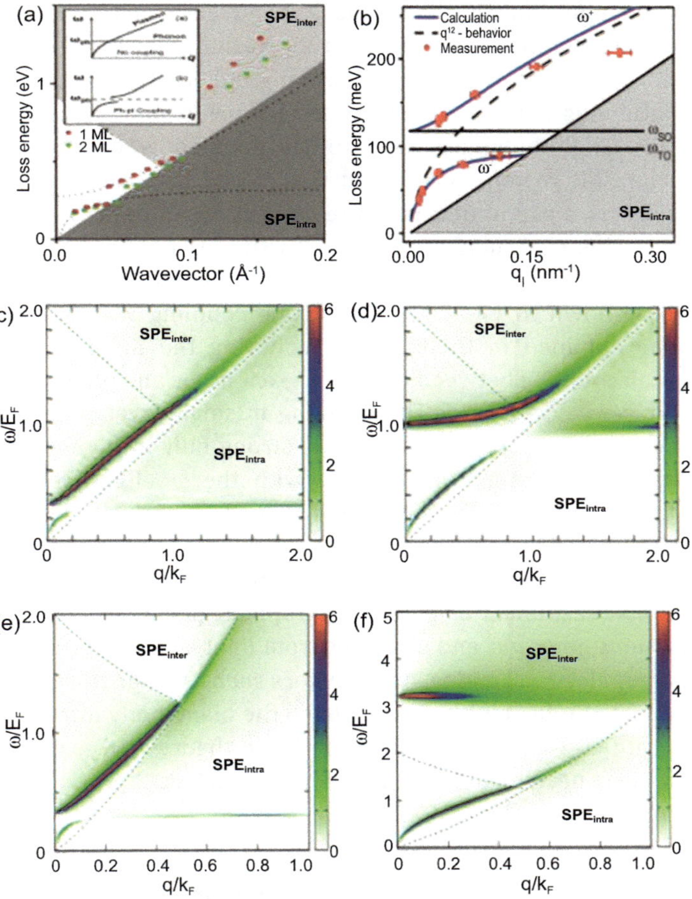

Figure 7.2 Dispersion of the plasmon-phonon coupled modes. (a) The energy loss peaks from angle-resolved EELS experiments for single-layer (1 ML) and bilayer (2 ML) graphene. Inset: schematic of the coupling. Reprinted with permission from Ref. 42, Copyright 2010 by the American Physical Society. (b) Dispersion of the coupled plasmon-phonon modes with a gap between $\omega^{\pm}$ modes. The red dots with error bars are experimental data. Reprinted with permission from Ref. 43, Copyright 2010 by the American Physical Society. (c–f) The strength of plasmon-phonon coupled modes in single-layer graphene based on polar substrate for electron concentration of 1013 cm^{-2} (c) and 1012 cm^{-2} (d), and in bilayer graphene for electron concentration of 1013 cm^{-2} (e) and 1012 cm^{-2} (f), respectively. Figures (c–f) reprinted with permission from Ref. 44, Copyright 2010 by the American Physical Society.

phonon modes in SiC cause the π^* and electrons in graphene oscillating. Moreover, a transition from plasmon-like to phonon-like dispersion is obtained with increasing graphene layers, where the discontinuous dispersions of $\omega^{\pm}$ modes are exhibited, as shown in Fig. 7.2. Both modes are strongly damped when they enter into the SPE regions. Combined with EELS data and numerical calculations, a gap in dispersion relation is found between the two modes, where ω^{+} modes converge to the LO phonon dispersion line and ω^{-} modes converge to the TO phonon dispersion line [43], as shown in Fig. 7.2b. The strong coupling with a gap can also be obtained in theory by considering the nonperturbative Coulomb coupling between electronic excitations and phonons [44], as shown in Figs. 7.2c–f. The coupling in single-layer graphene is strong at all densities; however, it is strong only at high densities for bilayer graphene, which agree with the $\omega^{\pm}$ EELS results in Ref. [42]. Consequently, substrates indeed impact the plasmon–phonon coupling seriously [45].

Anyway, plasmon–phonon coupling should be responsible for the observed quasi-linear dispersion of graphene plasmons. In addition to phonons, photons, and electrons from light, substrates, chemical doping or electrical gating, other factors such as magnetic field may also affect graphene plasmons through the many-body interactions. There are still many ambiguities in basic theory, and experiment phenomena need to be investigated in this subject.

7.3 Excitation of Surface Plasmons in Graphene

7.3.1 Spectroscopy Investigations of Surface Plasmons

Recent optical experiments have revealed the great prospects of plasmons in graphene for photonic device applications in the infrared (IR) and terahertz frequency ranges [47, 48]. Graphene can be used to build plasmonic metamaterials. Different from the conventional 2D electron gas system, a prominent room-temperature THz absorption peak is observed in graphene micro-ribbons array. The width of the ribbon will affect the results dramatically. Stacked structures can enhance the resonance further,

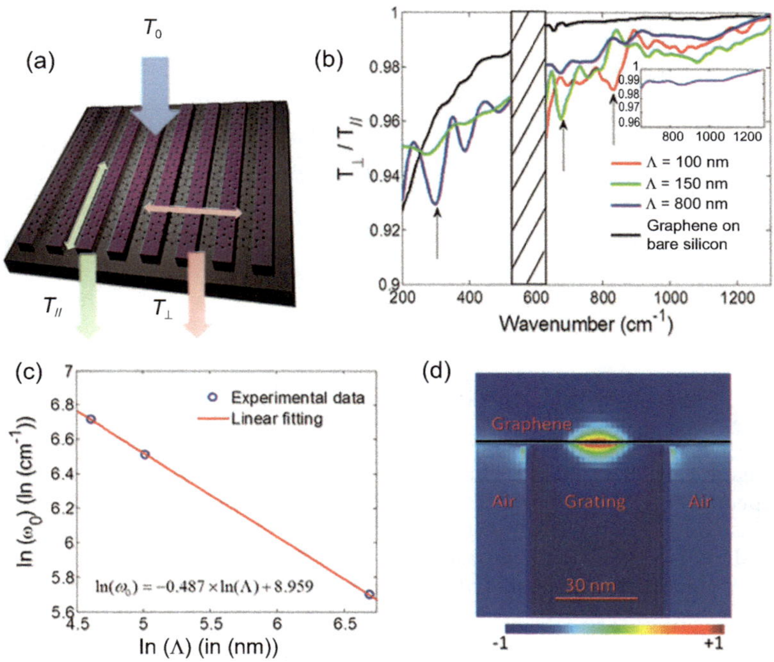

Figure 7.3 FTIR measurement results of fabricated guided-wave resonance graphene plasmonic devices. (a) Schematics of the normal-incidence transmission measurement and definitions of different polarizations for $T_\perp$ and $T_\parallel$. (b) FTIR transmission spectra in the far-infrared and mid-infrared for gratings with different periods. The arrows mark the positions of the resonances. Inset: The transmission spectrum of grating period $\Lambda = 800$ nm in mid-infrared region without graphene normalized to that of a bare silicon wafer. (c) The logarithm of resonance frequency ω_0 versus the logarithm of grating period Λ. (d) Simulated mode profiles using the finite difference time domain (FDTD) method at the resonance frequency. Reprinted with permission from Ref. 49, Copyright 2013, American Chemical Society.

and the transmittance can be improved by few-layer stacked graphene micro-disks.

Gao et al. demonstrated the excitation and gate control of highly confined SSP propagating through monolayer graphene using a silicon diffractive grating [49] (Fig. 7.3). The normal-incidence infrared transmission spectra exhibit pronounced dips due to guided-wave resonances, whose frequencies can be tuned over a

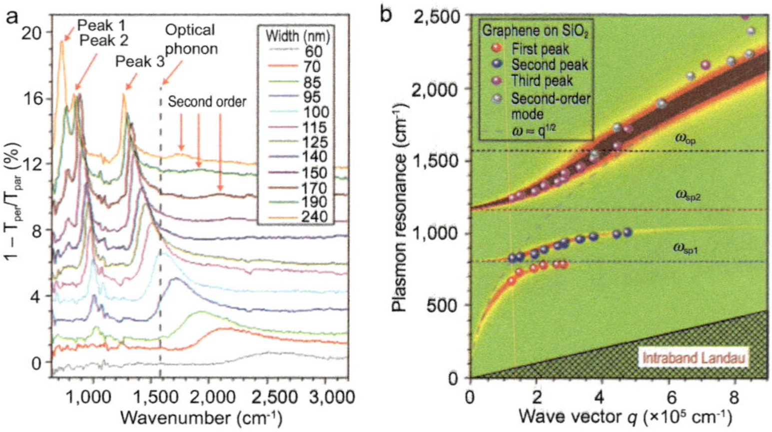

Figure 7.4 Plasmons in graphene nanoribbons on SiO_2. (a) Extinction spectra of graphene ribbons on SiO_2 with different ribbon widths. (b) Plasmon frequency as a function of wave vector $q = \pi / W_e$ for peaks 1, 2, and 3 and a second-order dipolar mode peak in (a). Reprinted by permission from Macmillan Publishers Ltd: *Nature Photonics*, Ref. 50, Copyright 2013.

range of $\sim$80 cm^{-1} by applying a gate voltage. This novel structure provides a way to excite and actively control plasmonic waves in graphene and is thus an important building block of graphene plasmonic systems. Yan et al. demonstrated mid-infrared plasmons in graphene nanostructures with dimensions as small as 50 nm [50] (Fig. 7.4). They revealed damping channels via graphene-intrinsic optical phonons and scattering from the edges. Plasmon lifetimes of 20 fs or less are observed when damping via the emission of graphene optical phonons is allowed. Furthermore, surface polar phonons in the SiO_2 substrate under graphene nanostructures lead to a significantly modified plasmon dispersion and damping, in contrast to the case of a nonpolar diamond-like carbon substrate.

7.3.2 Real-Space Imaging and Detection of Surface Plasmons

Since graphene SPs are the collective oscillation of the electrons, the changes in the electronic structure can be detected by a variety of direct and indirect methods, including EELS [43, 51], angle-resolved

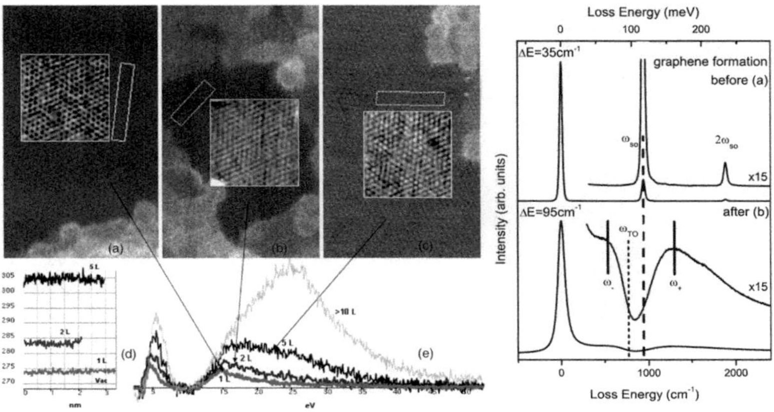

Figure 7.5 (a) Three upper panels are high-angle annular dark-field scanning transmission electron microscope images of one, two, and five layers of graphene; the two lower images show the intensity (left panel) along the long dimensions of the rectangular cyan-framed boxes and EELS spectra (right panel) of one, two, five, and several layers of graphene showing and plasmon in the three upper panels. Reprinted with permission from Ref. 51, Copyright 2008 by the American Physical Society. (b) Energy-loss spectra in specular direction for 6H-SiC(0001) before and after graphene formation. Reprinted with permission from Ref. 43, Copyright 2010 by the American Physical Society.

photoemission spectroscopy (ARPES) [52], and directly engineered light–plasmons coupling in a number of intriguing ways [47].

7.3.2.1 EELS and ARPES

The EELS provides a way to probe plasmons in graphene, by recording the energy loss of transmitted or reflected electrons. Using highly spatially resolved EELS experiment, π and $\pi + \sigma$ SP modes have been found in free-standing single graphene film at 4.7 and 14.6 eV, which are substantially redshifted from their values in graphite, and the integrated frequencies of energy peaks increase with the increase in thickness (see Fig. 7.5a) [51]. The energy-dependent energy-loss spectra also reveal the strong phonon–plasmon coupled modes in the graphene/silicon carbide heterosystem [43]. Two coupled phonon–plasmon modes $\omega_{\pm}$ can be observed in Fig. 7.5b.

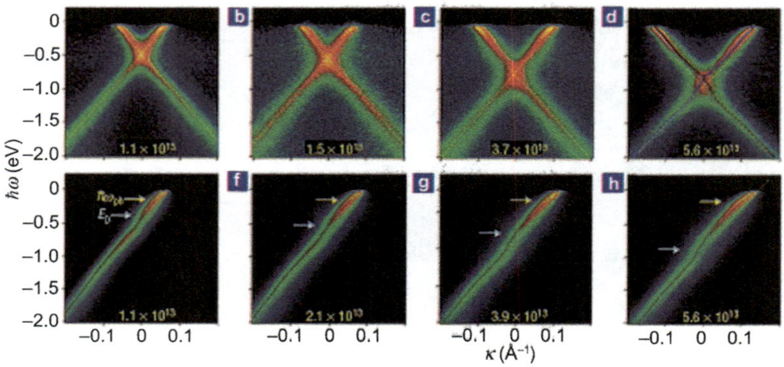

Figure 7.6 Experimental energy band by ARPES around the Dirac point of doped graphene based on SiC substrate, with different electron concentration (cm^{-2}). The kinked dispersion of the bands together with line width variations is clearly visible in the fitted peak positions (dotted lines). Reprinted with permission from Ref. 52, Copyright 2007, Nature Publishing Group.

The ARPES is also a powerful tool for the study of graphene plasmon. Bostwick et al. performed an ARPES measurement on a single-layer graphene, grown on the (001) surface of SiC. They found that the band structure around the *K* point is nonlinear or kinked, as shown in Fig. 7.6. The kinks in the bands' dispersion are caused by electron–plasmon interaction and the effect of phonons. Moreover, the kinked degree strongly depends on the electron concentration [52].

7.3.2.2 Light–plasmons coupling

The first look at light–plasmons coupling in graphene micro-ribbon arrays has been represented using polarized Fourier transform infrared spectroscopy [47]. The top and side views of a typical device incorporating periodic graphene micro-ribbon arrays are shown in Fig. 7.7a. In Fig. 7.7b, the absorption peak originating from plasmon oscillation can be observed in the extinction spectrum with perpendicular polarization, and plasmon resonance can be tuned in situ using an ion-gel top gate.

Furthermore, such plasmon excitation can be controlled by tailoring the width of the micro-ribbon (see Fig. 7.7c). When the dimension of the nanostructures is decreased to as small as 50 nm, the graphene plasmon resonance can be tuned into the mid-infrared regime (4–15 μm) [50]. Moreover, the lifetime of mid-infrared plasmon in graphene nano-ribbon strongly depends on the damping via interacting with substrate phonons and intrinsic graphene optical phonons. The study of the graphene plasmon in micro- and nano-ribbon arrays paves the way for applications of graphene in plasmonic waveguides, modulators, and detectors from terahertz to mid-infrared regimes. Graphene micro-ribbon based plasmonic waveguides can support two types of SP modes: the waveguiding modes, with the field concentration within the ribbon area, and the edge modes, with the field concentrated on the rims of the ribbons [53]. Recently, the hybridization in individual and paired graphene nano-ribbon has been demonstrated [54], and graphene nano-ribbon based waveguide and Y-shaped waveguide can be achieved by a proper design of nonuniform conductivity patterns across a flake of graphene [55].

Compared with graphene micro- and nano-ribbon, light–plasmons coupling in graphene micro- and nano-disk has no polarization dependence, which is desirable in applications such as detectors and filters [9, 10]. For graphene/insulator stacks array, the lateral disk–disk interaction can be neglected. As shown in Fig. 7.7d, the resonance frequency redshifts and peak intensity increases significantly with an increasing number of graphene layers; the peak intensity can reach ~50% for the five layers stacked structure [48]. Moreover, the plasmon can be tuned through changing the disk diameter, number of packed micro-disks, filling factor, and total carrier density. It was found that the graphene/insulator stacks can be used as an electromagnetic radiation shield with 97.5% effectiveness, a tunable far-infrared notch filter with 8.2 dB rejection ratio, and a tunable terahertz linear polarizer with 9.5 dB extinction ratio [48]. Besides, the electrical tunability and hybridization of localized plasmons in graphene nano-disks and nano-rings had been demonstrated [56]. The graphene plasmon could reach the near-infrared at 2.8 μm wavelength by engineering the graphene nano-ring pattern with ~20 nm spatial resolution. Such work illustrates

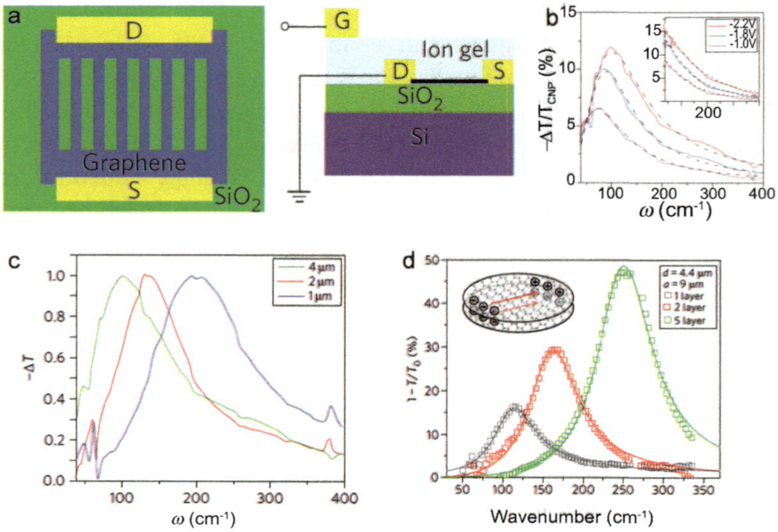

Figure 7.7 (a) Top view and side view of a typical device incorporating the graphene micro-ribbon array on an Si/SiO_2 substrate. The carrier concentration in graphene is controlled using the ion-gel top gate. (b) Control of terahertz resonance of plasmon excitations through electrical gating. Terahertz radiation was polarized perpendicular to the graphene ribbons (GRs). The plasmon resonance shifts to higher energy and gains oscillator strength with increased carrier concentration. For comparison, the inset shows corresponding spectra due to free carrier absorption for terahertz radiation polarized parallel to the ribbons. (c) Transmission spectra for different graphene micro-ribbon widths at the same doping concentration; the plasmon resonance clearly shifts to higher energy with decreasing ribbon widths. Figures (a–c) reprinted by permission from Macmillan Publishers Ltd: *Nature Nanotechnology*, Ref. 47, Copyright 2011. (d) The extinction spectra for microdisk arrays with one, two, and five graphene layers. Inset: schematic of disk–disk coupling in two closely stacked graphene disks. Reprinted by permission from Macmillan Publishers Ltd: *Nature Nanotechnology*, Ref. 48, Copyright 2012.

that the optical response of graphene can be brought closer to near-infrared region of interest for light modulation, switching, and sensing [56].

Recently, scattering-type scanning near-field optical microscopy (s-SNOM) has been applied to studying graphene SP [39, 57–

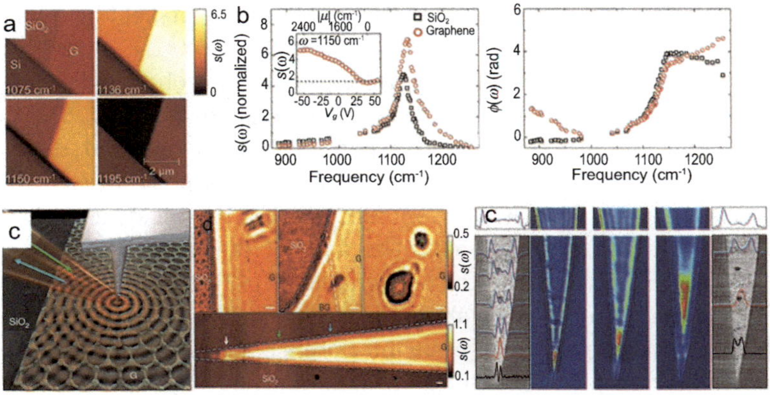

Figure 7.8 (a) Infrared near-field images displayed at four representative frequencies. The strong IR contrast between Si, SiO_2, and graphene (G) is clearly seen to vary systematically with the probing frequency. (b) Spectra of the near-field amplitude $s(\omega)$ and phase $\phi(\omega)$ extracted from images for SiO_2 (black squares) and graphene on SiO_2 (red circles). Figures (a,b) reprinted with permission from Ref. 57, Copyright 2011, American Chemical Society. (c) Diagram of an infrared nano-imaging experiment at the surface of graphene (G) on SiO_2. Concentric red circles illustrate plasmon waves launched by the illuminated tip. (d) Images of infrared amplitude s. These images show a characteristic interference pattern close to graphene edges and defects, and at the boundary between single (G) and bilayer (BG) graphene. (e) Images of a tapered graphene ribbon; the graphene plasmons strongly depend on dielectric properties of substrate and the excitation wavelength. Figures (c–e) reprinted by permission from Macmillan Publishers Ltd: *Nature*, Ref. 58, Copyright 2012.

62]. It is a nondestructive optical imaging technique with highly spatial resolution (~10 nm), and this technique compensates the momentum mismatch between graphene plasmon and light. In 2011, the infrared nanoscopy of Dirac plasmons at the graphene–SiO_2 interface was reported by mapping with SNOM for the first time (Fig. 7.8a) [57]. In Fig. 7.8b, graphene strongly enhances the amplitude $s(\omega)$ in the 1110–1250 cm^{-1} spectral region and also blueshifts the peak frequency by about 10 cm^{-1}. Both effects relate to the high carrier density in graphene. This can be verified by gating experiment in the inset. In addition, the steep increase in the phase below 970 cm^{-1} stems from plasmon–phonon near-field coupling [57].

Perhaps the most striking were the recent observations of graphene plasmons propagating in the real-space using SNOM [39, 58]. Figure 7.8c displays schematics of the infrared nano-imaging experiment. The beam of an infrared laser is focused on a sharp tip of an AFM. The tip concentrates the incident optical field at its apex to nano-focus. Because of the optical near-field interaction between the tip and graphene [63], the tip-scattered light contains information about the graphene SP. In Figs. 7.8d,e, the periodic fringe patterns are the interference between the launched and reflected plasmons. Infrared nano-imaging revealed that the extracted plasmon wavelength is more than 40 times smaller than the excitation wavelength and can be tuned by gating or changing the infrared light wavelength. The strong confinement and tunable property of graphene plasmon makes it be an ideal medium for the design and miniaturization of nano-optoelectronic devices and functionalities.

Graphene-propagating plasmons can be reflected and scattered by graphene edges [39, 58, 62], nanoscale morphological defects [39], graphene grain boundaries [59], and nanometer-size terraces [60]. The grain boundaries in a large-area CVD graphene film can be observed by mapping scanning plasmon interferometry images using SNOM (see Fig. 7.9a). The visualization and investigation of graphene grain boundaries using SNOM uncover a microscopic mechanism that is responsible for the low electron mobility observed in CVD-grown graphene [59]. The scanning plasmon interferometry images also reveal the strong plasmons reflection at nanometer-size terraces in monolayer graphene (see Fig. 7.9b). When the height exceeds a certain height about 1.5 nm, the reflection signal will increase by more than 20% of its value at graphene edges, which even approaches 50% when the height is around 5 nm. So the extremely compact nanostructure can be used to tailor the propagation in graphene-based circuits [60].

Due to the unique properties of graphene plasmon, including high confinement, good tunability, and low loss, it can be a promising plasmonic material alternative to the noble metals for various

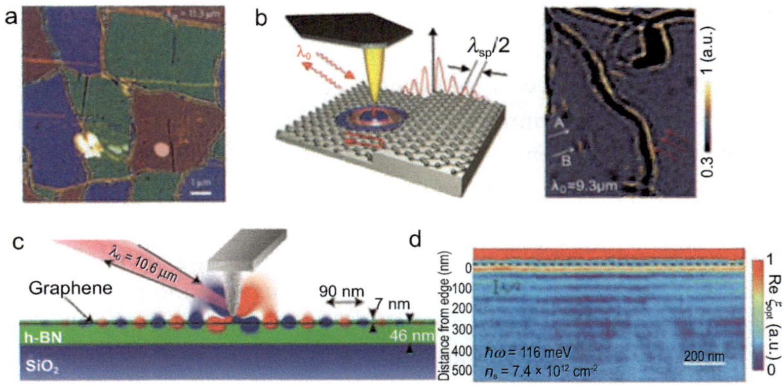

Figure 7.9 (a) A larger-area scan of a typical sample revealing multiple grains (displayed with different false colors) defined by the twin fringes due to grain boundaries and grain overlaps. Reprinted by permission from Macmillan Publishers Ltd: *Nature Nanotechnology*, Ref. 59, Copyright 2013. (b) (Left) Schematics of near-filed nano-imaging the graphene plasmons on SiC terraces; (right) optical near-field amplitude of quasi-free-standing monolayer epitaxial graphene on a 6H–SiC substrate. Red arrows indicate fringes along a substrate step. Reprinted with permission from Ref. 60, Copyright 2013, American Chemical Society. (c) Simplified side-view schematic of the s-SNOM measurement. The red and blue show the simulated in-plane component of the electric field of a dipole source oscillating at a photon energy of 116 meV coupling to graphene plasmons. (d) s-SNOM optical signal from a 2D scan of the tip position, near the graphene edge (dashed line) at room temperature. Figures (c,d) reprinted by permission from Macmillan Publishers Ltd: *Nature Materials*, Ref. 61, Copyright 2015.

applications. But a strong plasmon damping has been observed in graphene, because of impurity scattering and many-body effects. Recently, highly confined low-loss plasmon has been found in a new type of heterostructure, as shown in Fig. 7.9c [61], graphene encapsulated between two films of hexagonal boron nitride (h-BN). The near-field optical images show the propagating plasmons in this heterostructure (see Fig. 7.9d) and reveal unprecedentedly low plasmon damping in such plasmonic medium.

7.4 Plasmonic Devices

The SP has been both verified by infrared transmission spectroscopy and observed by scanning near-field optical microscopy in real-space [39, 50]. Recently, a large deal of work has been devoted to finding materials with improved plasmonic performance and fast tunability, including the use of super-conductors in the sub-THz regime and conductive oxides in the visible and near-infrared. Graphene is emerging as a very promising plasmonic material due to its carrier density tunability and high carrier mobility. A typical device of periodic graphene micro-ribbon arrays is shown in Fig. 7.10.

In order to achieve light–plasmons coupling, Ju et al. explored plasmon excitations by engineering graphene film into micro-ribbon arrays and found that graphene plasmon resonances can be tuned over a broad terahertz frequency range by changing micro-ribbon width and in situ electrostatic doping. However, it is noteworthy

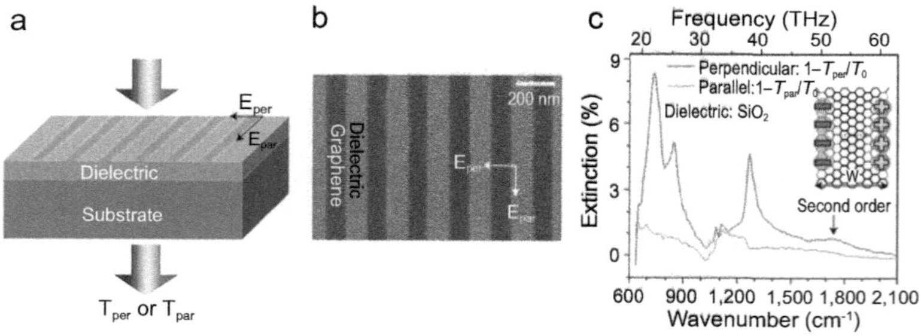

Figure 7.10 Schematics of the experiment. (a) Mid-infrared transmission measurement scheme for graphene nanoribbons. The excitation light is broadband. False colors are used for excitation light (rainbow) and on-resonance light in graphene nanoribbons (red). (b) SEM image of a typical array of graphene nanoribbons. The width of the ribbon is 100 nm. (c) Extinction spectra of a ribbon array on SiO_2 with the incident light polarization perpendicular (red) and parallel (grey) to the ribbons. Ribbon width is 240 nm. A weak second-order mode is indicated. Inset: dipole oscillation in a graphene ribbon. Reprinted by permission from Macmillan Publishers Ltd: *Nature Photonics*, Ref. 50, Copyright 2013.

that the dependence of both the plasmonic resonance frequency and magnitude on carrier concentration is relatively weak in single-layer graphene.

Due to the flexible 2D nature, SPs in graphene can perform versatile properties by various geometry and topography. Many different properties of SPs in graphene were exploited in different structures such as multilayer, micro/nano-ribbon, micro/nano-disk/antidot, ring, and stacks or even corrugation, making plasmons in graphene a very promising field in both semi-classical and quantum frameworks. Anyway, with different structures, graphene can be regarded as a versatile metamaterial for transmission, absorption, modulator, polarizer, or even for the mysterious cloaking [64].

Generally, with the scale of graphene above few dozens of nanometers, the semi-classical theory (e.g., Maxwell equations) is sufficient enough to describe the performance of plasmons. In a word, in addition to the single-layer sheet, other graphene structures are also potential platforms for plasmons. For the separated single-layer graphene arrays or few-layer graphene, SPs are still dominating due to the ultrathin thickness, although the bulk plasmons cannot be ignored any more.

7.4.1 Surface Plasmons in Graphene

It is known that layer graphene is a promising material for many potential applications, including plasmonics. The approaches for studying plasmons in single-layer graphene can be adapted to describe the plasmons in bi- and multilayer graphene. However, the properties of graphene with different layers will be distinguished from each other because of the different band structures and many-body interactions. SPs in these ultrathin sheets are still the dominating components, and they can propagate along the sheets in the double-layer structure. It is worth mentioning that SPs in the graphene sheets usually couple with each other by the symmetric and antisymmetric modes where the velocity and damping can be easily controlled by gate voltage [65, 66].

7.4.2 Surface Plasmons in Graphene Ribbons

Graphene can be easily tailored into various geometries for practical applications, in which SPs will be different too. The most investigated structures are GRs. By reducing the degree of freedom, SPPs in these ribbons can propagate in the wanted direction, while LSPR is enhanced due to confinement in other directions. As a result, GRs can be used as waveguides. Waveguides and edge modes can be found in the THz frequency range when SPs propagate along GRs, which are separated from each other by a gap in wave numbers. Moreover, higher frequency or wider ribbon can increase the number of SP modes, and the propagation length, rather than wave vector, is strongly sensitive to the relaxation time of charge carriers. The LSPR in GRs can also enhance the optical absorption.

For incident EM wave polarized perpendicular to GRs, the prominent room-temperature optical absorption peaks can be obtained in the THz region and the resonances can also be tuned by electrical doping, incident angle, and the array scales. By suppressing transmission in these ribbons, light passing through GR arrays can even be completely absorbed [67]. Although being approximately proportional to the coverage of graphene, the absorption in GRs arrays is still stronger than that in the continuous graphene sheet due to the sufficiently high relaxation time.

7.4.3 Surface Plasmons in Graphene Disks, Rings

To reduce the dimensionality further, graphene with structures of disks, rings is studied [68–71]. The EM field in graphene antidot can also be regarded as a dipole, while plasmons in graphene ring can be treated as the plasmons hybridization from a graphene disk. For similar sizes, the order of energies of the plasmons hybridization for different graphene structures from low to high is as follows: ring (symmetric coupling), disk, antidote, and ring (antisymmetric coupling), respectively. The enhancement factor of the EM field by plasmons in graphene ring (symmetric coupling) can reach as large as 10^3 times in the THz region, which is almost 20 times larger than a similar structure made of gold [70]. Furthermore, the relatively high relaxation times of the charge carriers in these graphene structures

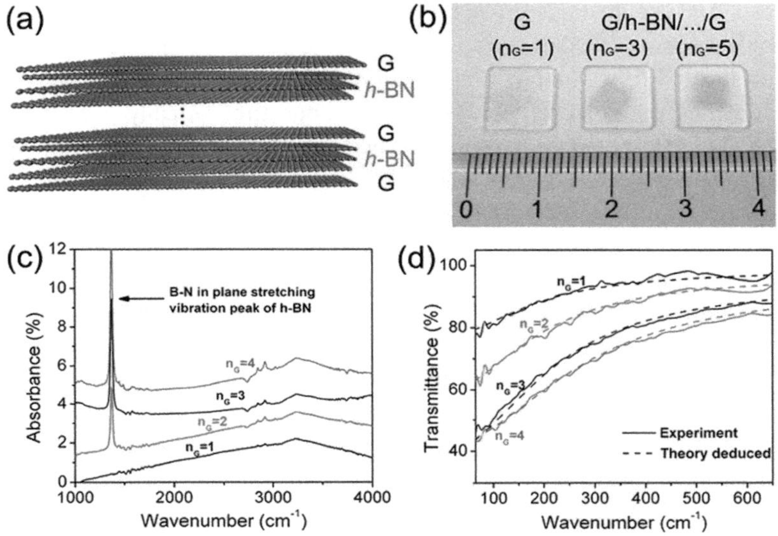

Figure 7.11 (a) Scheme showing layer-by-layer stacking of the graphene and h-BN. (b) Graphene/h-BN stacks with different layer numbers transferred on 1 cm square quartz substrates. (c,d) Mid-IR and far-IR transmittance of the graphene/h-BN heterostructures. The sharp peak around 1370 cm^{-1} corresponds to BN in plane stretching vibration of h-BN. Reprinted from Ref. 76 with permission from John Wiley and Sons.

enhance the coupling between plasmons and other quasi-particles and lead to enhanced absorption and suppressed transmission in the THz region.

Large-scale patterns with graphene micro-/nanostructures are necessary for practical applications. The SP coupling between graphene nanostructures on the same plane is relatively weak; however, it is strong in the stacked structures [71].

7.4.4 Surface Plasmons in Graphene/Insulator or TI Stacks

The surfaces of topological insulators (TIs) and graphene have very similar low-energy electronic structures in which the conduction and valence bands touch at Dirac points. Dirac fermions also occur in the 2D electron gas that forms at the surface of TIs as a result of the strong spin–orbit interaction existing in the insulating bulk phase [72]. The collective modes, or so-called spin-plasmons, originated from the fact that density fluctuations induce transverse spin fluctuations and vice versa have been theoretically predicted in TIs [73]. Pietro et al. reported the first experimental evidence of Dirac plasmon excitations in a TI (i.e., Bi_2Se_3) with thin micro-ribbon arrays of different widths [74]. The plasmonic resonance in TI is attributed to the bounded oscillation of Dirac quasi-particles of the conducting 2D edge state. According to theoretical studies [54], the plasmon in graphene will interact and hybridize strongly with itself by inserting a thin dielectric spacer between two layers of graphene to form a sandwich structure like "metal–dielectric–metal." An intriguing issue is whether the interaction and hybridization are observable between graphene and TIs considering the fact that the plasmonic waves in these two layered materials have similar origin [75].

Yan et al. fabricated graphene/insulator multilayer stacks to effectively enhance the plasmonic resonance frequency and magnitude [48]. It is interesting to find that the peak conductivity at the plasmonic resonance increases linearly with layer number, which is a direct consequence of the unique carrier density scaling law of the plasmonic resonance of Dirac fermions. As shown in Fig. 7.11, Zhang et al. used atomically thin h-BN films as spacers to enable strong plasmon coupling between adjacent graphene layers in vertical stacks, which produces 2.1 times upshift of plasmonic resonance frequency and 3 times enhancement of its magnitude in graphene/h-BN stacks [76].

7.5 Conclusion

Apart from conventional materials, the study of 2D materials has emerged to be a significant field of study for a variety of applications. Graphene plasmonics are also coming of age and show very attractive features, which include extremely high localization, strong confinement, efficient and strong light–matter interactions, relatively large lifetimes, tunability, and electric controllability. Hopefully, light manipulation with intrinsic graphene plasmons and the accessibility of quantum optical regimes promise a revolution in light creation, routing, modulation, computing, and detection.

References

1. Polman, A. (2008). Plasmonics applied, *Science*, **322**, pp. 868–869.
2. Liu, Z., Lee, H., Xiong, Y., Sun, C., and Zhang, X. (2007). Far-field optical hyperlens magnifying sub-diffraction-limited objects, *Science*, **315**, pp. 1686.
3. Hunt, L. B. (1976). The true story of Purple of Cassius, *Gold Bull*, **9**, pp. 134–139.
4. Rodríguez-Lorenzo, L., Álvarez-Puebla, R. A., Pastoriza-Santos, I., Mazzucco, S., Stéphan, O., Kociak, M., Liz-Marzán, L. M., and García de Abajo, F. J. (2009). Zeptomol detection through controlled ultrasensitive surface-enhanced Raman scattering, *Journal of the American Chemical Society*, **131**, pp. 4616–4618.
5. Grzelczak, M., Perez-Juste, J., Mulvaney, P., and Liz-Marzan, L. M. (2008). Shape control in gold nanoparticle synthesis, *Chemical Society Reviews*, **37**, pp. 1783–1791.
6. Halas, N. J., Lal, S., Chang, W.-S., Link, S., and Nordlander, P. (2011). Plasmons in strongly coupled metallic nanostructures, *Chemical Reviews*, **111**, pp. 3913–3961.
7. Kneipp, K., Wang, Y., Kneipp, H., Perelman, L. T., Itzkan, I., Dasari, R. R., and Feld, M. S. (1997). Single molecule detection using surface-enhanced raman scattering (SERS), *Physical Review Letters*, **78**, pp. 1667–1670.
8. Atwater, H. A., and Polman, A. (2010). Plasmonics for improved photovoltaic devices, *Nature Materials*, **9**, pp. 205–213.
9. Knight, M. W., Sobhani, H., Nordlander, P., and Halas, N. J. (2011). Photodetection with active optical antennas, *Science*, **332**, pp. 702–704.
10. Loo, C., Lowery, A., Halas, N., West, J., and Drezek, R. (2005). Immunotargeted nanoshells for integrated cancer imaging and therapy, *Nano Letters*, **5**, pp. 709–711.
11. Harutyunyan, H., Palomba, S., Renger, J., Quidant, R., and Novotny, L. (2010). Nonlinear dark-field microscopy, *Nano Letters*, **10**, pp. 5076–5079.
12. Wood, R. (1902). XLII. On a remarkable case of uneven distribution of light in a diffraction grating spectrum, *The London, Edinburgh, and*

Dublin Philosophical Magazine and Journal of Science, **4**, pp. 396–402.

13. Fano, U. (1941). The theory of anomalous diffraction gratings and of quasi-stationary waves on metallic surfaces (Sommerfeld's waves), *Journal of the Optical Society of America*, **31**, pp. 213–222.
14. Rayleigh, L. (1907). III. Note on the remarkable case of diffraction spectra described by Prof. Wood, *The London, Edinburgh, and Dublin Philosophical Magazine and Journal of Science*, **14**, pp. 60–65.
15. Crookes, W. (1879). On radiant matter, *Journal of the Franklin Institute*, **108**, pp. 305–316.
16. Langmuir, I. (1928). Oscillations in ionized gases, *Proceedings of the National Academy of Sciences*, **14**, pp. 627–637.
17. Tonks, L., and Langmuir, I. (1929). Oscillations in ionized gases, *Physical Review*, **33**, pp. 195–210.
18. Bohm, D., and Pines, D. (1951). A collective description of electron interactions. I. Magnetic interactions, *Physical Review*, **82**, pp. 625–634.
19. Ritchie, R. H. (1957). Plasma losses by fast electrons in thin films, *Physical Review*, **106**, pp. 874–881.
20. Stern, E. A., and Ferrell, R. A. (1960). Surface plasma oscillations of a degenerate electron gas, *Physical Review*, **120**, pp. 130–136.
21. Ritchie, R. H., and Wilems, R. E. (1969). Photon-plasmon interaction in a nonuniform electron gas. I, *Physical Review*, **178**, pp. 372–381.
22. Bardeen, J., and Pines, D. (1955). Electron-phonon interaction in metals, *Physical Review*, **99**, pp. 1140–1150.
23. Pitarke, J. M., Silkin, V. M., Chulkov, E. V., and Echenique, P. M. (2007). Theory of surface plasmons and surface-plasmon polaritons, *Reports on Progress in Physics*, **70**, pp. 1.
24. William, L. B. (2006). Surface plasmon–polariton length scales: A route to sub-wavelength optics, *Journal of Optics A: Pure and Applied Optics*, **8**, pp. S87.
25. Murray, W. A., and Barnes, W. L. (2007). Plasmonic materials, *Advanced Materials*, **19**, pp. 3771–3782.
26. Raether, H. (1988). Surface plasmons on smooth surfaces. In *Surface Plasmons on Smooth and Rough Surfaces and on Gratings*. Springer-Verlag Berlin Heidelberg.
27. Kreibig, U., and Vollmer, M. (1995). *Optical Properties of Metal Clusters*. Springer-Verlag Berlin Heidelberg.
28. Pendry, J. B. (2000). Negative refraction makes a perfect lens, *Physical Review Letters*, **85**, pp. 3966.

29. Zheludev, N. I., and Kivshar, Y. S. (2012). From metamaterials to metadevices, *Nature Materials*, **11**, pp. 917–924.

30. Atwater, H. A., and Polman, A. (2010). Plasmonics for improved photovoltaic devices, *Nature Materials*, **9**, pp. 205–213.

31. Anker, J. N., Hall, W. P., Lyandres, O., Shah, N. C., Zhao, J., and Van Duyne, R. P. (2008). Biosensing with plasmonic nanosensors, *Nature Materials*, **7**, pp. 442–453.

32. Novoselov, K. S., Geim, A. K., Morozov, S., Jiang, D., Zhang, Y., Dubonos, S., Grigorieva, I., and Firsov, A. (2004). Electric field effect in atomically thin carbon films, *Science*, **306**, pp. 666–669.

33. Grigorenko, A., Polini, M., and Novoselov, K. (2012). Graphene plasmonics, *Nature Photonics*, **6**, pp. 749–758.

34. Neto, A. C., Guinea, F., Peres, N., Novoselov, K. S., and Geim, A. K. (2009). The electronic properties of graphene, *Reviews of Modern Physics*, **81**, pp. 109.

35. Koppens, F. H., Chang, D. E., and Garcia de Abajo, F. J. (2011). Graphene plasmonics: A platform for strong light–matter interactions, *Nano Letters*, **11**, pp. 3370–3377.

36. Giuliani, G., and Vignale, G. (2005). *Quantum Theory of the Electron Liquid*. Cambridge university press.

37. Gass, M. H., Bangert, U., Bleloch, A. L., Wang, P., Nair, R. R., and Geim, A. (2008). Free-standing graphene at atomic resolution, *Nature Nanotechnology*, **3**, pp. 676–681.

38. Barnes, W. L., Dereux, A., and Ebbesen, T. W. (2003). Surface plasmon subwavelength optics, *Nature*, **424**, pp. 824–830.

39. Fei, Z., Rodin, A., Andreev, G., Bao, W., McLeod, A., Wagner, M., Zhang, L., Zhao, Z., Thiemens, M., and Dominguez, G. (2012). Gate-tuning of graphene plasmons revealed by infrared nano-imaging, *Nature*, **487**, pp. 82–85.

40. Kohn, W. (1959). Image of the fermi surface in the vibration spectrum of a metal, *Physical Review Letters*, **2**, pp. 393.

41. Tse, W.-K., Hu, B. Y.-K., and Sarma, S. D. (2008). Chirality-induced dynamic Kohn anomalies in graphene, *Physical Review Letters*, **101**, pp. 066401.

42. Liu, Y., and Willis, R. F. (2010). Plasmon-phonon strongly coupled mode in epitaxial graphene, *Physical Review B*, **81**, pp. 081406.

43. Koch, R., Seyller, T., and Schaefer, J. (2010). Strong phonon–plasmon coupled modes in the graphene/silicon carbide heterosystem, *Physical Review B*, **82**, pp. 201413.

44. Hwang, E., Sensarma, R., and Sarma, S. D. (2010). Plasmon–phonon coupling in graphene, *Physical Review B*, **82**, pp. 195406.

45. Ong, Z.-Y., and Fischetti, M. V. (2012). Theory of interfacial plasmon–phonon scattering in supported graphene, *Physical Review B*, **86**, pp. 165422.

46. Sensale-Rodriguez, B., Yan, R., Kelly, M. M., Fang, T., Tahy, K., Hwang, W. S., Jena, D., Liu, L., and Xing, H. G. (2012). Broadband graphene terahertz modulators enabled by intraband transitions, *Nature Communications*, **3**, pp. 780.

47. Ju, L., Geng, B., Horng, J., Girit, C., Martin, M., Hao, Z., Bechtel, H. A., Liang, X., Zettl, A., and Shen, Y. R. (2011). Graphene plasmonics for tunable terahertz metamaterials, *Nature Nanotechnology*, **6**, pp. 630–634.

48. Yan, H., Li, X., Chandra, B., Tulevski, G., Wu, Y., Freitag, M., Zhu, W., Avouris, P., and Xia, F. (2012). Tunable infrared plasmonic devices using graphene/insulator stacks, *Nature Nanotechnology*, **7**, pp. 330–334.

49. Gao, W., Shi, G., Jin, Z., Shu, J., Zhang, Q., Vajtai, R., Ajayan, P. M., Kono, J., and Xu, Q. (2013). Excitation and active control of propagating surface plasmon polaritons in graphene, *Nano Letters*, **13**, pp. 3698–3702.

50. Yan, H., Low, T., Zhu, W., Wu, Y., Freitag, M., Li, X., Guinea, F., Avouris, P., and Xia, F. (2013). Damping pathways of mid-infrared plasmons in graphene nanostructures, *Nature Photonics*, **7**, pp. 394–399.

51. Eberlein, T., Bangert, U., Nair, R., Jones, R., Gass, M., Bleloch, A., Novoselov, K., Geim, A., and Briddon, P. (2008). Plasmon spectroscopy of free-standing graphene films, *Physical Review B*, **77**, pp. 233406.

52. Bostwick, A., Ohta, T., Seyller, T., Horn, K., and Rotenberg, E. (2007). Quasiparticle dynamics in graphene, *Nature Physics*, **3**, pp. 36–40.

53. Nikitin, A. Y., Guinea, F., García-Vidal, F., and Martín-Moreno, L. (2011). Edge and waveguide terahertz surface plasmon modes in graphene microribbons, *Physical Review B*, **84**, pp. 161407.

54. Christensen, J., Manjavacas, A., Thongrattanasiri, S., Koppens, F. H., and García de Abajo, F. J. (2011). Graphene plasmon waveguiding and hybridization in individual and paired nanoribbons, *ACS Nano*, **6**, pp. 431–440.

55. Vakil, A., and Engheta, N. (2011). Transformation optics using graphene, *Science*, **332**, pp. 1291–1294.

56. Fang, Z., Thongrattanasiri, S., Schlather, A., Liu, Z., Ma, L., Wang, Y., Ajayan, P. M., Nordlander, P., Halas, N. J., and García de Abajo, F. J. (2013). Gated tunability and hybridization of localized plasmons in nanostructured graphene, *ACS Nano*, **7**, pp. 2388–2395.

57. Fei, Z., Andreev, G. O., Bao, W., Zhang, L. M. S., McLeod, A., Wang, C., Stewart, M. K., Zhao, Z., Dominguez, G., and Thiemens, M. (2011). Infrared nanoscopy of Dirac plasmons at the graphene–SiO_2 interface, *Nano Letters*, **11**, pp. 4701–4705.
58. Chen, J., Badioli, M., Alonso-González, P., Thongrattanasiri, S., Huth, F., Osmond, J., Spasenović, M., Centeno, A., Pesquera, A., and Godignon, P. (2012). Optical nano-imaging of gate-tunable graphene plasmons, *Nature*, **487**, pp. 77–81.
59. Fei, Z., Rodin, A., Gannett, W., Dai, S., Regan, W., Wagner, M., Liu, M., McLeod, A., Dominguez, G., and Thiemens, M. (2013). Electronic and plasmonic phenomena at graphene grain boundaries, *Nature Nanotechnology*, **8**, pp. 821–825.
60. Chen, J., Nesterov, M. L., Nikitin, A. Y., Thongrattanasiri, S., Alonso-Gonzaílez, P., Slipchenko, T. M., Speck, F., Ostler, M., Seyller, T., and Crassee, I. (2013). Strong plasmon reflection at nanometer-size gaps in monolayer graphene on SiC, *Nano Letters*, **13**, pp. 6210–6215.
61. Woessner, A., Lundeberg, M. B., Gao, Y., Principi, A., Alonso-González, P., Carrega, M., Watanabe, K., Taniguchi, T., Vignale, G., Polini, M., Hone, J., Hillenbrand, R., and Koppens, F. H. L. (2015). Highly confined low-loss plasmons in graphene–boron nitride heterostructures, *Nature Materials*, **14**, pp. 421–425.
62. Gerber, J. A., Berweger, S., O'Callahan, B. T., and Raschke, M. B. (2014). Phase-resolved surface plasmon interferometry of graphene, *Physical Review Letters*, **113**, pp. 055502.
63. Cvitkovic, A., Ocelic, N., and Hillenbrand, R. (2007). Analytical model for quantitative prediction of material contrasts in scattering-type near-field optical microscopy, *Optics Express*, **15**, pp. 8550–8565.
64. Chen, P.-Y., and Alù, A. (2011). Atomically thin surface cloak using graphene monolayers, *ACS Nano*, **5**, pp. 5855–5863.
65. Wang, B., Zhang, X., Yuan, X., and Teng, J. (2012). Optical coupling of surface plasmons between graphene sheets, *Applied Physics Letters*, **100**, pp. 131111.
66. Svintsov, D., Vyurkov, V., Ryzhii, V., and Otsuji, T. (2013). Voltage-controlled surface plasmon-polaritons in double graphene layer structures, *Journal of Applied Physics*, **113**, pp. 053701.
67. Alaee, R., Farhat, M., Rockstuhl, C., and Lederer, F. (2012). A perfect absorber made of a graphene micro-ribbon metamaterial, *Optics Express*, **20**, pp. 28017–28024.

68. Nikitin, A. Y., Guinea, F., and Martin-Moreno, L. (2012). Resonant plasmonic effects in periodic graphene antidot arrays, *Applied Physics Letters*, **101**, pp. 151119.

69. Wang, W. (2012). Plasmons and optical excitations in graphene rings, *Journal of Physics: Condensed Matter*, **24**, pp. 402202.

70. Liu, P., Cai, W., Wang, L., Zhang, X., and Xu, J. (2012). Tunable terahertz optical antennas based on graphene ring structures, *Applied Physics Letters*, **100**, pp. 153111.

71. Yan, H., Xia, F., Li, Z., and Avouris, P. (2012). Plasmonics of coupled graphene micro-structures, *New Journal of Physics*, **14**, pp. 125001.

72. Zhang, J., Triola, C., and Rossi, E. (2014). Proximity effect in graphene–topological-insulator heterostructures, *Physical Review Letters*, **112**, pp. 096802.

73. Raghu, S., Chung, S. B., Qi, X.-L., and Zhang, S.-C. (2010). Collective modes of a helical liquid, *Physical Review Letters*, **104**, pp. 116401.

74. Di Pietro, P., Ortolani, M., Limaj, O., Di Gaspare, A., Giliberti, V., Giorgianni, F., Brahlek, M., Bansal, N., Koirala, N., and Oh, S. (2013). Observation of Dirac plasmons in a topological insulator, *Nature Nanotechnology*, **8**, pp. 556–560.

75. Stauber, T. (2014). Plasmonics in Dirac systems: From graphene to topological insulators, *Journal of Physics: Condensed Matter*, **26**, pp. 123201.

76. Zhang, K., Yap, F. L., Li, K., Ng, C. T., Li, L. J., and Loh, K. P. (2014). Large scale graphene/hexagonal boron nitride heterostructure for tunable plasmonics, *Advanced Functional Materials*, **24**, pp. 731–738.

Chapter 8

Graphene as Optical Limiters

Hui Ying Hoh[a] **and Qiaoliang Bao**[a,b]

[a]*Department of Materials Science and Engineering, Monash University, 22 Alliance Lane, Building 69, Clayton, 3800, Victoria, Australia*
[b]*Institute of Functional Nano & Soft Materials (FUNSOM), Soochow University, 199 Ren-Ai Road, Suzhou Industrial Park, Suzhou Jiangsu 215123, China*
qiaoliang.bao@gmail.com

8.1 Introduction

Optical limiters are materials or devices with high transmittance for incident light with low power or intensity, such as ambient light, but low transmittance for input with high power or intensity, which is potentially damaging. The most important function of such materials is in the protection of optical sensors, especially the human eye, although they are also useful in other applications such as modifying pulse shape of an optical signal [1]. Interest in such materials grew as high-power laser systems are increasingly being employed for scientific, military, and commercial applications.

Active optical limiters usually use a shutter that blocks out the potentially dangerous beam. Since it requires feedback from

Graphene Photonics, Optoelectronics, and Plasmonics
Edited by Qiaoliang Bao, Hui Ying Hoh, and Yupeng Zhang

ISBN 978-981-4745-60-4 (Hardcover), 978-1-315-19667-1 (eBook)
www.panstanford.com

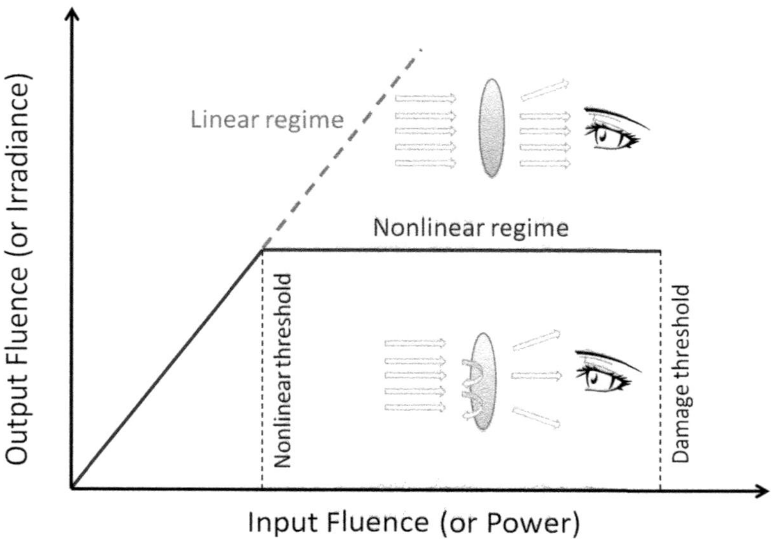

Figure 8.1 Ideal behavior of a nonlinear optical limiter. The dynamic range is defined as the ratio of the transmittance at the nonlinear threshold to the transmittance at the damage threshold.

various components of the system, its operation is complex and is not practical for protection against fast lasers. Passive optical limiters, on the other hand, rely on a nonlinear optical material to moderate the transmittance of the incident light as a function of its intensity.

Generally, an ideal optical limiter should fulfil three criteria, as depicted in Fig. 8.1. First, it should have a low nonlinear (fluence) threshold and large dynamic range, since the main purpose of an optical limiter is sensor protection and the extent of damage is highly dependent on the output fluence. Second, it should be able to respond instantaneously to a range of lasers, be it a continuous wave or femtosecond pulses. Finally, the optical limiter should have a broad spectral response, operating effectively over a wide wavelength range, typically from the visible to infrared region [2]. In reality, however, no material could satisfy all the qualities, and the selection of a material (or hybrid system) is based often on the required application and cost.

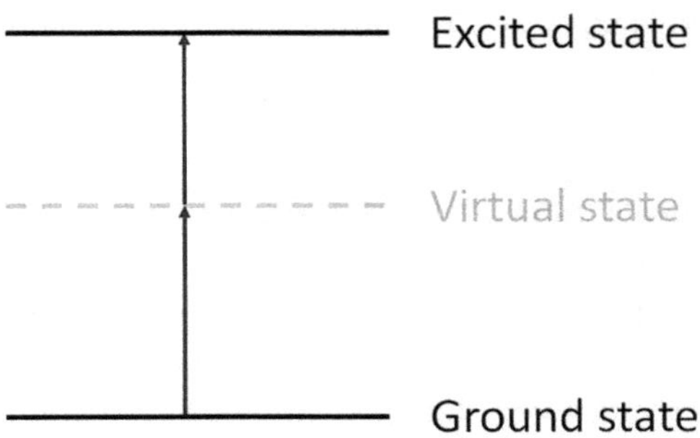

Figure 8.2 Two-photon absorption.

8.2 Nonlinear Effect

Two main mechanisms are available for the nonlinear effect. The first is nonlinear absorption, which can be further explained by two different mechanisms: multiphoton absorption and excited-state absorption. The second is nonlinear scattering involving microplasmas and microbubbles.

8.2.1 Nonlinear Absorption—Multiphoton Absorption

Multiphoton absorption involves the transition from the ground state to an excited state through the absorption of two or more photons. In two-photon absorption, for example, the first photon excites the atom or molecule to a virtual state between the ground and excited states and the second photon brings it to the final excited state (Fig. 8.2). The strength of absorption is highly dependent on the light intensity. This mechanism was first studied in inorganic materials, but later organic molecules with long conjugation system, analogous to a large antenna, and those with substitution by strong donor and acceptor groups were also considered.

8.2.2 Nonlinear Absorption—Excited-State Absorption (Reverse Saturable Absorption)

Excited-state absorption usually occurs when the intermediate states of a system have a longer lifetime than the time between two consecutive pulses. It is a cumulative process in which electrons are first promoted to lower excited states through interaction with an initial pulse, followed by absorption of subsequent pulses to populate higher energy states. Therefore, the fluence-integrated flux over time, instead of the intensity of the incident light, is the determining factor for the strength of absorption. In reverse saturable absorption, the absorption cross section of the excited state is larger than that of the ground state. This usually occurs in organic molecules that demonstrate intersystem crossing to the triplet state [1].

8.2.3 Nonlinear Scattering

In solution-based optical limiters, the limiting effects typically arise from nonlinear scattering. Thermally induced nonlinear scattering involves the formation and growth of scattering centers. Nucleation of the scattering centers occurs when an incident laser pulse causes heating of suspended particles, leading to vaporization and ionization of the particles, which are known as microplasmas. Microbubbles are then produced in the solvent due to the thermal energy released by the expanding microplasmas. These microplasmas and microbubbles strongly scatter light from the transmitted direction, resulting in a nonlinear decrease in the energy. Such a process occurs in the nanosecond time scale and is, therefore, only applicable when the pulse laser operates in the nanosecond regimes.

8.3 Carbon-Based Optical Limiters

Due to similarities in chemical structure, it is interesting to look at other forms of carbon-based materials that have been extensively investigated prior to graphene. Carbon black suspensions, owing to their low-cost and simple preparation, were the first to be considered. Later, research efforts shifted to suspensions containing fullerenes and carbon nanotubes. The limiting properties of carbon black and carbon nanotube suspensions arise from nonlinear scattering, whereas that of fullerenes may exhibit significant excited-state absorption [3–5]. Chemical modification or blending could also allow for other nonlinear effects, improving the overall optical limiting performance.

For example, Izard et al. suspended grafted single-wall carbon nanotubes in modified Stilbene-3 dye solution, a multiphoton absorber [6]. Due to the cumulative effect of multiphoton absorption and scattering, this blend exhibits a better limiting effect than pure carbon nanotubes for laser pulses operating in the nanosecond range. The authors also noted that the choice of chromophore, as well as the concentration of the blend, is critical to the cumulative effect. As shown in Fig. 8.3, the cumulative effect is more pronounced

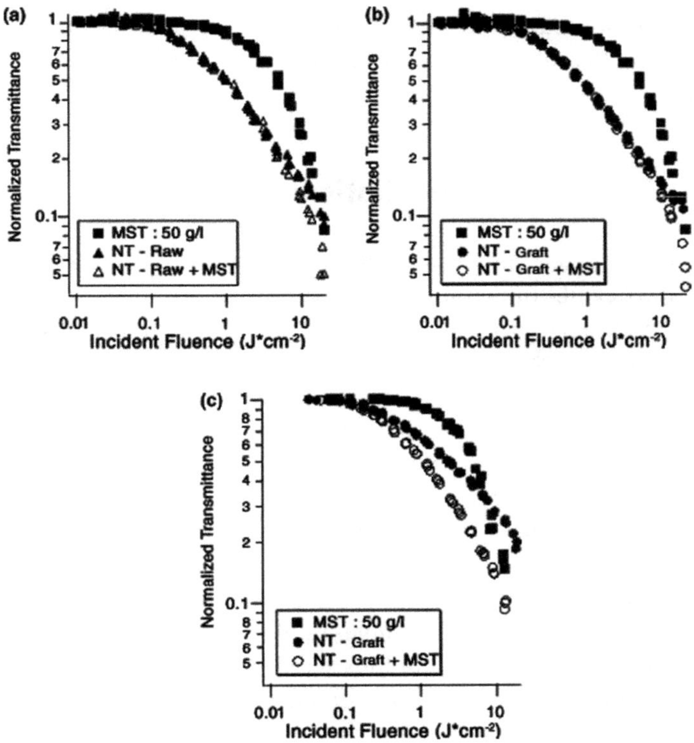

Figure 8.3 Optical limiting results at 532 nm, for 3 ns pulses of raw carbon nanotubes (a) and grafted nanotubes at regular and low concentrations (70% and 82% transmittance) in (b) and (c), respectively. MST: modified Stilbene-3; NT: nanotube. Reprinted from Ref. 6, Copyright 2004, with permission from Elsevier.

at lower concentrations (82% transmittance compared to 70% transmittance). It was concluded that when the efficiency of the chromophore is comparable to that of the carbon nanotubes, the resulting blend will demonstrate the most improvement in optical limiting performance. More importantly, since multiphoton absorption comes into play, the blend may be useful for sub-nanosecond pulses, lending promise for optical limiting in broad temporal and spectral ranges [2].

Readers interested in carbon-based optical limiters are encouraged to refer to the reviews by Chin et al., Sun et al., and Vivien et al. for more information [2, 7, 8]. For solid state systems based on fullerenes and carbon nanotubes, readers may also refer to the review by Brusatin and Signorini, as well as that by Chen et al. [9, 10].

8.4 Graphene-Based Optical Limiters

Despite immense progress in the development of optical limiters, materials exhibiting nonlinear absorption do not display practical limiting property over the entire visible and near-infrared ranges. This is because the ratio of the excited-state to ground-state cross-sections is highly wavelength dependent. Carbon black and carbon nanotube suspensions utilizing nonlinear scattering are often used in broadband optical limiting applications [5, 7]. However, as mentioned above, these suspensions cannot be used for laser pulses shorter than a few nanoseconds, for instance in ultrafast spectroscopy. In addition, such suspensions require replenishment in solvent systems due to the breakdown of material, and the formation of black spots or pinholes renders them ineffective in solid composite films. These limitations set the motivation for looking at graphene-based materials as broadband optical limiters for ultrafast lasers.

8.4.1 Graphene Dispersions

In pure graphene, the excited-state electrons have an ultrafast (femtosecond) relaxation timescale, and this is not desirable for optical limiting applications. Most studies on the optical limiting properties of graphene were performed using solution-dispersed graphene oxide derivatives, in which the limiting effect was attributed to nonlinear scattering. One such study was performed by Wang et al. in 2009 [11]. Graphene dispersions in various solvents were considered. The selected solvents—N-methyl-2-pyrrolidone (NMP), N,N-dimethylacetamide (DMA), and γ-butyrolactone (GBL)—have similar surface energies as graphite (around 70–80 mJ/m^2), reducing the energy required to overcome the van der Waals forces between two graphene sheets, and thus enabling stable dispersions containing single- or a few-layered graphene sheets. As depicted in Fig. 8.4a, all dispersions demonstrated gradually reduced transmissions with increasing incident energy at both 532 nm and 1064 nm, indicating a broadband limiting effect.

The authors also compared the optical limiting behavior of graphene with single-walled carbon nanotubes (SWNTs) and C_{60}. As seen in Fig. 8.5, the demonstrated optical limiting performances of the three carbon-based materials are comparable at 532 nm. Graphene also exhibits optical limiting response at both 532 nm and 1064 nm, suggesting its potential in broadband limiting applications.

In cases where charge transfer occurs between graphene and other components, the optical limiting effect may also arise from nonlinear absorption, as found in a study by Xu et al. [12]. Optoelectronically active porphyrin molecules (TPP-NH_2) were covalently bonded to graphene oxide, as shown in Fig. 8.6a. The hybrid (TPP-NHCO-SPFGraphene) is highly soluble and stable in organic solvents. More importantly, in open-aperture Z-scan studies, TPP-NHCO-SPFGraphene displayed enhanced optical limiting properties (Fig. 8.6b). TPP-NHCO-SPFGraphene outperforms not only its individual components (TPP-NH_2 and graphene oxide) and a control blend sample, a mixture of TPP-NH_2 and graphene oxide in a 1:1 ratio, but also the benchmark material, C_{60}. The optical limiting behavior of the hybrid arises from a combination of photoinduced electron and/or energy transfer, nonlinear absorption, and nonlinear scattering. Therefore, consideration of photoinduced electron and/or energy transfer is imperative when selecting a hybrid material for optical limiting purposes.

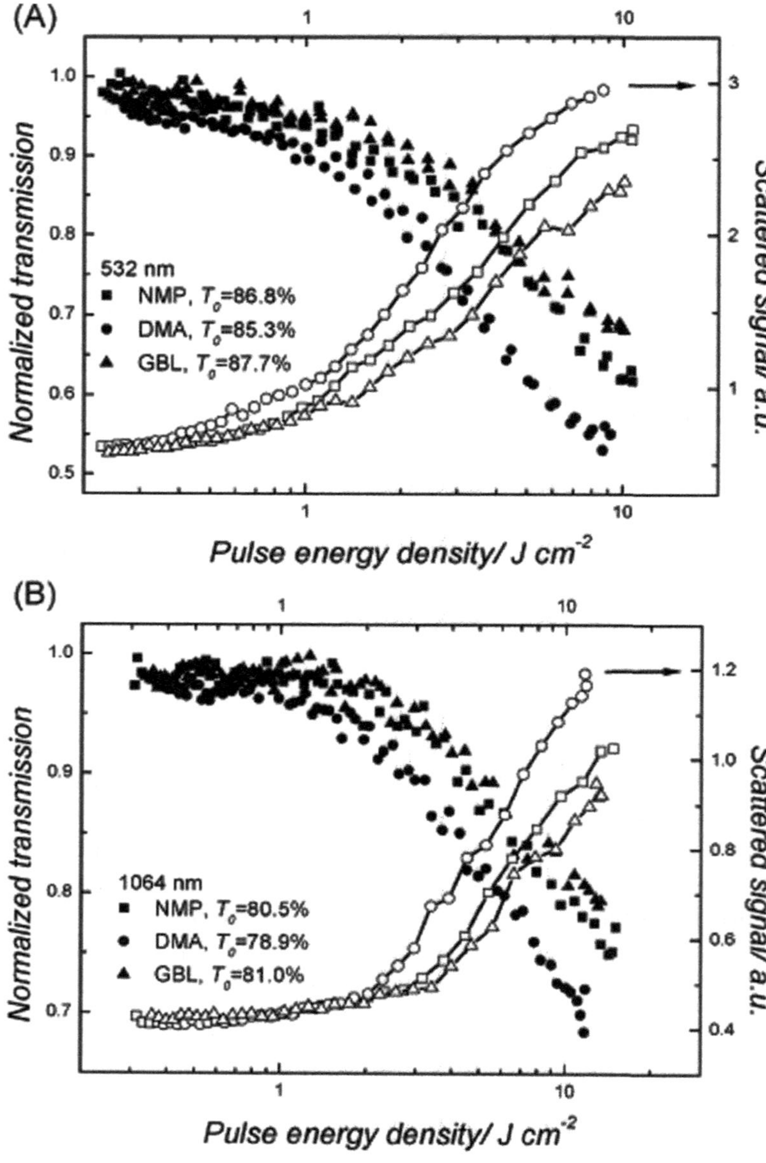

Figure 8.4 Nonlinear transmission and scattering of graphene in NMP, DMA, and GBL for excitation at (a) 532 nm and (b) 1064 nm. The linear transmission T_0 is also presented. Reprinted from Ref. 11 with permission from John Wiley and Sons.

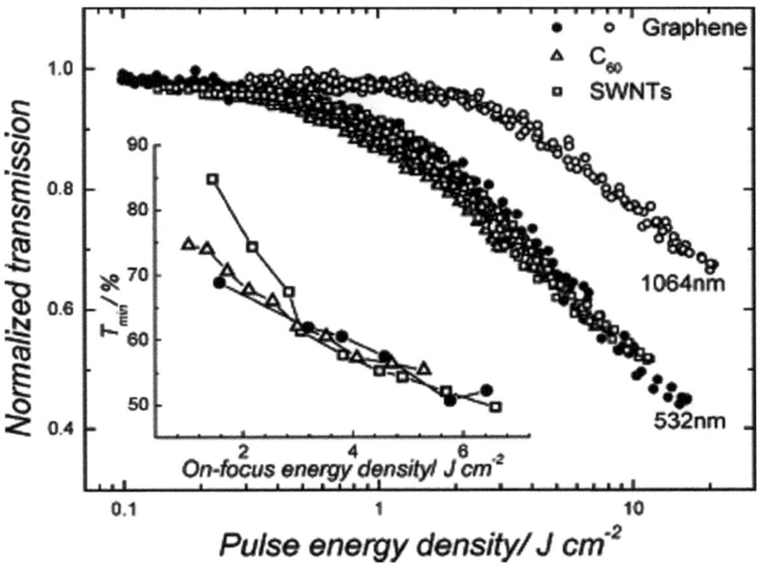

Figure 8.5 Optical limiting behaviour of graphene, C_{60}, and SWNTs at ~82% transmittance. Inset: minimal transmission as a function of on-focus energy density. Reprinted from Ref. 11 with permission from John Wiley and Sons.

Later, it was discovered that the length of laser pulses also affects the optical limiting effect of graphene oxide derivatives: In ultrashort laser pulses (femtosecond), multiphoton absorption dominates, whereas in the case of longer pulses (nanosecond), excited-state absorption plays a significant role [13, 14]. This scenario is particularly relevant when the relaxation dynamics of graphene are altered due to the modification of its electronic structure through chemical functionalization. For instance, small π-conjugated domains embedded within an insulating matrix, constructed from alkyl-grafted, sub-stoichiometric graphene oxide (sub-GO_x), permit the existence and confinement of long-lived electron–hole pairs, which are not isolated from (instead interact strongly with) the surrounding environment (Fig. 8.7a)[15]. As a result, a strong matrix effect was observed. A new benchmark for highly effective optical limitation across visible and near-infrared spectral regions was set, at pulse fluences well below the onset of microbubble/microplasma formation, with an onset of 10 mJ/cm

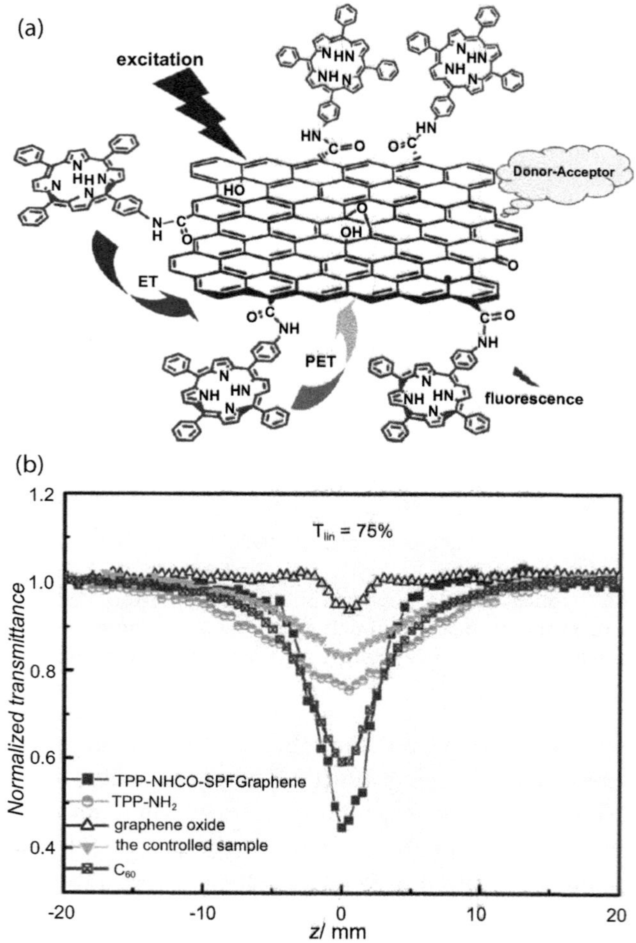

Figure 8.6 Porphyrin–graphene nanohybrid. (a) Schematic representation of part of the structure of the hybrid (TPP-NHCO-SPFGraphene). ET: energy transfer; PET: photoinduced electron transfer. (b) Open-aperture Z-scan results with 532 nm pulses. Reprinted from Ref. 12 with permission from John Wiley and Sons.

for a linear transmittance of 70%, as shown in Fig. 8.7b. This "giant optical limiting effect" is observed in both solvent systems and polymer films, and it is worth noting that the material breakdown

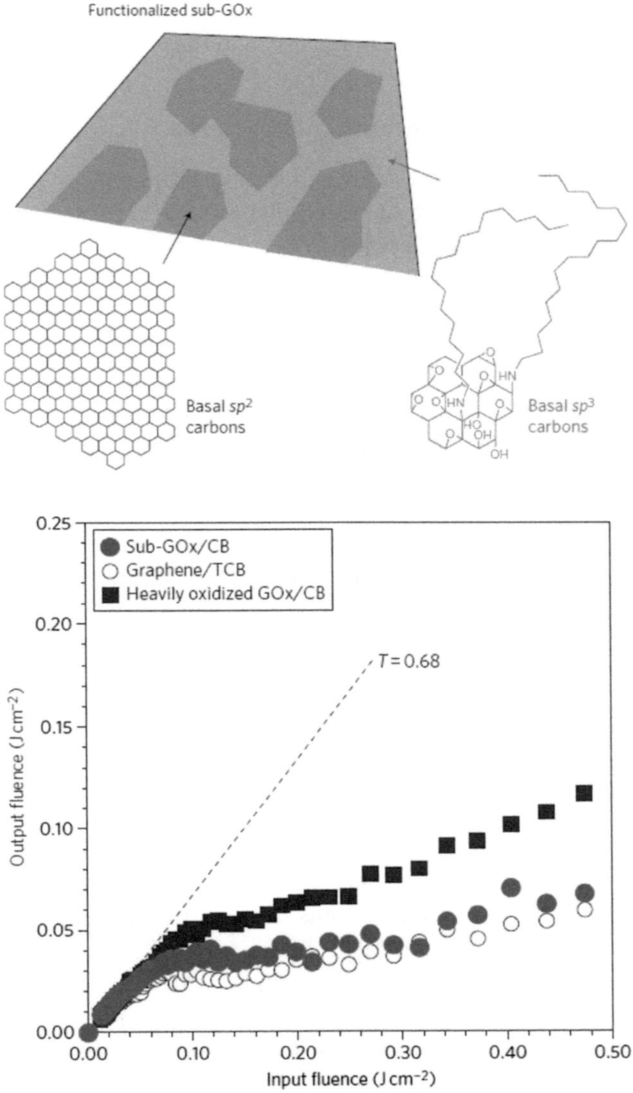

Figure 8.7 (Top) Schematic structure of sub-stoichiometric graphene oxide (sub-GO_x). (Bottom) Comparing the nonlinear transmittance of sub-GO_x, ultrasonically exfoliated graphene, and heavily oxidized GO_x. CB: chlorobenzene; TCB: 1,2,4-trichlorobenzene. Reprinted by permission from Macmillan Publishers Ltd: *Nature Photonics*, Ref. 15, Copyright 2011.

that occurs at high fluences in carbon black suspensions, carbon nanotubes, or graphite was not observed in these graphene samples.

8.4.2 Composites Containing Graphene

Follow-up studies focus on incorporating graphene or graphene oxides in an assortment of host polymers to harness functional matrix effects, for instance polyvinyl alcohol, to offer a low-cost, non-toxic solution [16], or polyimide, to improve the overall nonlinear optical performance through charge transfer from photoexcited polyimide to graphene [17]. Other works, such as the ones by Ouyang et al., include the addition of inorganic nanomaterials (CdS, MoS_2, AgS_2, ZnO) to the graphene–polymer composite to enhance the excited-state absorption, attributed to synergetic effects between graphene and the inorganic material [18–21]. For example, in a composite containing graphene and zinc oxide nanoparticles (ZnO), interfacial charge transfer between graphene and ZnO results in a two-photon process in addition to the ones involving only graphene or only ZnO (Fig. 8.8) [21]. This additional process, in turn, is the source of enhanced nonlinear absorption in the composite.

Another class of host materials to be considered includes silica glasses, which, compared to polymers, have higher melting point and excellent transparency in the UV region, thus making them more viable options for optical applications. Having incorporated graphene oxide sheets in silica glass homogenously, Tao et al. observed an onset of nonlinear optical response at a low input fluence of 1.3 J/cm^2 (532 nm), thus demonstrating the potential of such glasses as optical limiters [22].

Optical limiting properties of graphene can also be tailored by altering the morphology of graphene. Xu et al. reported a simple and reversible method to control the curling and unfolding, and hence the optoelectronic properties of graphene oxide through interactions with porphyrin molecules [23]. The strain induced by porphyrin interactions on the basal planes of the graphene oxide created a resulting hybrid with an obvious curled morphology, and interestingly, the hybrid with curved morphology displayed superior optical limiting properties relative to the planar hybrid.

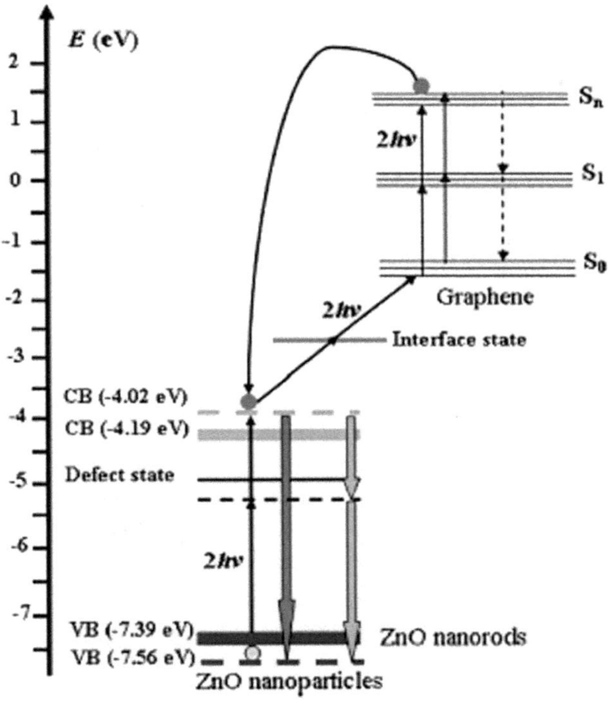

Figure 8.8 Energy-level diagram of graphene/ZnO composites, showing the two-photon processes. Reprinted from Ref. 21, Copyright 2014, with permission from Elsevier.

8.5 Conclusion

In this chapter, we outlined the role of optical limiters as well as the nonlinear effect. Works on the optical limiting behavior of various carbon materials, namely carbon black suspensions, fullerenes, and carbon nanotubes, were summarized, after which important studies on graphene were highlighted. As advancements in growth, dispersion, and modification techniques of graphene continue, novel designs and applications are constantly being explored. Recently, Zheng et al. reported that the nonlinear refractive index and nonlinear absorption coefficient in high-quality graphene or graphene oxide films can be tuned over large dynamic ranges with controllable laser fluence [24]. Excluding previous studies that involved graphene in solutions and composites, this is the first to consider third-order nonlinear effects in graphene films [25]. The third-order nonlinearity, in particular the nonlinear refraction (Kerr effect), is a critical consideration for applications involving ultrafast lasers since it can be considered a quasi-instantaneous process [26]. Exploitation of different nonlinearities may be the solution to high-speed, broadband optical limiting; and graphene, with its unique properties, will not only be regarded as a new benchmark material, but will also see uses in wide ranging functions.

References

1. Tutt, L. W., and Boggess, T. F. (1993). A review of optical limiting mechanisms and devices using organics, fullerenes, semiconductors and other materials, *Prog Quantum Electron*, **17**, pp. 299–338.
2. Chin, K. C., Gohel, A., Elim, H. I., Chen, W., Ji, W., Chong, G. L., Sow, C. H., and Wee, A. T. S. (2006). Modified carbon nanotubes as broadband optical limiting nanomaterials, *J Mater Res*, **21**, pp. 2758–2766.
3. Jin, M. H. C., Durstock, M., and Dai, L. (2006). Optical limiters and photovoltaic devices based on C60, carbon nanotubes and their nanocomposites, in *Carbon Nanotechnology* (Elsevier), pp. 611–631.
4. Tutt, L. W., and Kost, A. (1992). Optical limiting performance of C60 and C70 solutions, *Nature*, **356**, pp. 225–226.
5. Chen, P., Wu, X., Sun, X., Lin, J., Ji, W., and Tan, K. L. (1999). Electronic structure and optical limiting behavior of carbon nanotubes, *Phys Rev Lett*, **82**, pp. 2548–2551.
6. Izard, N., Ménard, C., Riehl, D., Doris, E., Mioskowski, C., and Anglaret, E. (2004). Combination of carbon nanotubes and two-photon absorbers for broadband optical limiting, *Chem Phys Lett*, **391**, pp. 124–128.
7. Sun, Y. A. P., Riggs, J. E., Henbest, K. B., and Martin, R. B. (2000). Nanomaterials as optical limiters, *J Nonlinear Opt Phys Mater*, **9**, pp. 481–503.
8. Vivien, L., Lançon, P., Riehl, D., Hache, F., and Anglaret, E. (2002). Carbon nanotubes for optical limiting, *Carbon*, **40**, pp. 1789–1797.
9. Brusatin, G., and Signorini, R. (2002). Linear and nonlinear optical properties of fullerenes in solid state materials, *J Mater Chem*, **12**, pp. 1964–1977.

10. Chen, Y., Lin, Y., Liu, Y., Doyle, J., He, N., Zhuang, X., Bai, J., and Blau, W. J. (2007). Carbon nanotube-based functional materials for optical limiting, *J Nanosci Nanotechnol*, **7**, pp. 1268–1283.
11. Wang, J., Hernandez, Y., Lotya, M., Coleman, J. N., and Blau, W. J. (2009). Broadband nonlinear optical response of graphene dispersions, *Adv Mater*, **21**, pp. 2430–2435.
12. Xu, Y., Liu, Z., Zhang, X., Wang, Y., Tian, J., Huang, Y., Ma, Y., Zhang, X., and Chen, Y. (2009). A graphene hybrid material covalently functionalized with porphyrin: Synthesis and optical limiting property, *Adv Mater*, **21**, pp. 1275–1279.
13. Feng, M., Zhan, H., and Chen, Y. (2010). Nonlinear optical and optical limiting properties of graphene families, *Appl Phys Lett*, **96**, pp. 033107.
14. Liu, Z., Wang, Y., Zhang, X., Xu, Y., Chen, Y., and Tian, J. (2009). Nonlinear optical properties of graphene oxide in nanosecond and picosecond regimes, *Appl Phys Lett*, **94**, pp. 021902.
15. Lim, G. K., Chen, Z. L., Clark, J., Goh, R. G. S., Ng, W. H., Tan, H. W., Friend, R. H., Ho, P. K. H., and Chua, L. L. (2011). Giant broadband nonlinear optical absorption response in dispersed graphene single sheets, *Nat Photon*, **5**, pp. 554–560.
16. Husaini, S., Slagle, J. E., Murray, J. M., Guha, S., Gonzalez, L. P., and Bedford, R. G. (2013). Broadband saturable absorption and optical limiting in graphene-polymer composites, *Appl Phys Lett*, **102**, pp. 191112.
17. Gan, Y., Feng, M., and Zhan, H. (2014). Enhanced optical limiting effects of graphene materials in polyimide, *Appl Phys Lett*, **104**, pp. 171105.
18. Ouyang, Q., Yu, H., Wu, H., Lei, Z., Qi, L., and Chen, Y. (2013). Graphene/MoS_2 organic glasses: Fabrication and enhanced reverse saturable absorption properties, *Opt Mater*, **35**, pp. 2352–2356.
19. Ouyang, Q., Di, X., Lei, Z., Qi, L., Li, C., and Chen, Y. (2013). Enhanced reverse saturable absorption in graphene/Ag_2S organic glasses, *Phys Chem Chem Phys*, **15**, pp. 11048–11053.
20. Ouyang, Q., Yu, H., Xu, Z., Zhang, Y., Li, C., Qi, L., and Chen, Y. (2013). Synthesis and enhanced nonlinear optical properties of graphene/CdS organic glass, *Appl Phys Lett*, **102**, pp. 031912.
21. Ouyang, Q., Xu, Z., Lei, Z., Dong, H., Yu, H., Qi, L., Li, C., and Chen, Y. (2014). Enhanced nonlinear optical and optical limiting properties of graphene/ZnO hybrid organic glasses, *Carbon*, **67**, pp. 214–220.
22. Tao, L., Zhou, B., Bai, G., Wang, Y., Yu, S. F., Lau, S. P., Tsang, Y. H., Yao, J., and Xu, D. (2013). Fabrication of covalently functionalized graphene oxide

incorporated solid-state hybrid silica gel glasses and their improved nonlinear optical response, *J Phys Chem C*, **117**, pp. 23108–23116.

23. Xu, H., Wu, P., Liao, C., Lv, C., and Gu, Z. (2014). Controlling the morphology and optoelectronic properties of graphene hybrid materials by porphyrin interactions, *Chem Commun*, **50**, pp. 8951–8954.
24. Zheng, X., Jia, B., Chen, X., and Gu, M. (2014). In situ third-order non-linear responses during laser reduction of graphene oxide thin films towards on-chip non-linear photonic devices, *Adv Mater*, **26**, pp. 2699–2703.
25. Hendry, E., Hale, P. J., Moger, J., Savchenko, A. K., and Mikhailov, S. A. (2010). Coherent nonlinear optical response of graphene, *Phys Rev Lett*, **105**, pp. 097401.
26. Steinmeyer, G. (2006). Nonlinear and short pulse effects, in *Handbook of Optoelectronics*, eds. Dakin, J., and Brown, R. G. W. (New York: Taylor & Francis), pp. 307–327.

Chapter 9

Graphene-Based Light-Emitting Diodes

Jialu Zheng,[a] Hui Ying Hoh,[a] Yupeng Zhang,[a] and Qiaoliang Bao[a,b]

[a] *Department of Materials Science and Engineering, Monash University, 22 Alliance Lane, Building 69, Clayton, 3800, Victoria, Australia*
[b] *Institute of Functional Nano & Soft Materials (FUNSOM), Soochow University, 199 Ren-Ai Road, Suzhou Industrial Park, Suzhou Jiangsu 215123, China*
qiaoliang.bao@gmail.com

9.1 Introduction

A light-emitting diode (LED) is a solid state lighting source that can convert electrical energy into optical radiation [1, 2]. LEDs have been widely applied in many fields, including display backlight, communications, medical services, signage, and general illumination [3–7]. LEDs offer design flexibility, from zero-dimensional lighting to three-dimensional lighting, with one-dimensional lighting and two-dimensional lighting in between [2]. Compared with traditional lighting, LEDs provides higher performance, such as ultrahigh speed response (micro-second-level on-off switching), a wider range of controllable color temperatures (4500–12,000 K), a wider operating temperature range (−20°C to 85°C), no low-temperature startup problems, better mechanical impact resistance, and low UV radiation [7].

Graphene Photonics, Optoelectronics, and Plasmonics
Edited by Qiaoliang Bao, Hui Ying Hoh, and Yupeng Zhang

ISBN 978-981-4745-60-4 (Hardcover), 978-1-315-19667-1 (eBook)
www.panstanford.com

Graphene is a two-dimensional material which has attracted a lot of attention due to its excellent physical and chemical properties, including excellent electron mobility, extreme hardness, and good thermal and chemical stability [8, 9]. Originally, few-layers graphene was achieved through exfoliation from graphite by the scotch tape method. However, these films are of small areas and not suitable for application in LEDs. After a few years, solution processing of graphene was reported [10]. In this process, graphene thin films were prepared by the solution processing of functionalized graphene followed by high-temperature annealing to reduce graphene oxide into graphene. In recent years, another significant way of producing graphene is chemical vapor deposition (CVD) [11]. This method has become the most common way to prepare large-scale graphene thin films. The graphene produced by CVD has excellent electric conductivity. Since the graphene can be transferred to other substrates such as SiO_2/Si, it can be used as electrodes for photonic devices [2].

Traditionally, the electrodes material of LED is indium tin oxide (ITO) [12]. However, both indium and tin are limited resources, which increase the price of LEDs. The brittleness of ITO also renders it unsuitable for flexible devices. All these drawbacks of ITO limit the application of LED [12]. In this case, an alternative material is highly required for LED technology. Compared with ITO, graphene not only has excellent electrical conductivity, thermal and chemical stability, and near transparency, but more importantly a lower price, which is more economical. Figure 9.1 shows the schematic of directly growth graphene (DG) as electrode in GaN-based LEDs.

9.2 Development of Graphene-Based LEDs

9.2.1 Selective Doped Graphene as Anodic Electrode or Cathodic Electrode

Although ITO has been widely used as anode electrodes in the LED industry, the high price and brittleness of ITO limit its application. Due to the excellent properties, graphene has been selected as a potential alternative to ITO in LED [13–16]. However, compared

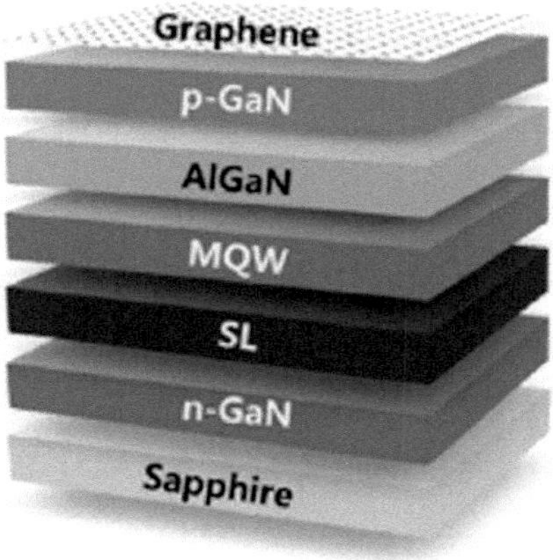

Figure 9.1 Schematic of the DG/LED structure. Reprinted with permission from Ref. 11, Copyright 2014, American Chemical Society.

with ITO, still some properties of graphene require to be improved. First, graphene has high sheet resistance ($>300\ \Omega^{-1}$) and low work function (WF) ($\sim$4.4 eV), while ITO has low sheet resistance ($10\ \Omega^{-1}$) and relatively high WF ($\sim 4.7 \leq \text{WF} \leq 4.9$ eV) [5]. The low WF of graphene causes high injection barrier between the graphene anode and the overlying organic layers, which makes the hole injection process unfavorable. As a result, graphene-based OLEDs have shown poorer current efficiencies (CEs, unit cd/A) than ITO-based devices [5, 14, 17]. In addition, the low conductivity of pristine graphene films results in high operating voltage, which limits the luminous efficiencies (LEs, unit lm/W) of the device [5].

In 2012, Han et al. demonstrated a method to increase the surface WF of graphene film to 5.95 eV and reduce the sheet resistance to $30\ \Omega^{-1}$ [13]. They used conducting polymer compositions to modify the graphene and dope it with p-dopants, HNO_3 or $AuCl_3$, which created a WF gradient from the graphene to the overlying organic layer. In addition, they also made a self-organized hole injection layer (HIL). This HIL is essential because of the huge hole injection energy barrier ($\sim$1.0 eV) between the graphene anode and the hole transport layer (HTL). Figure 9.2 illustrates how the HIL influences the hole injection process of the graphene-based OLED devices. The

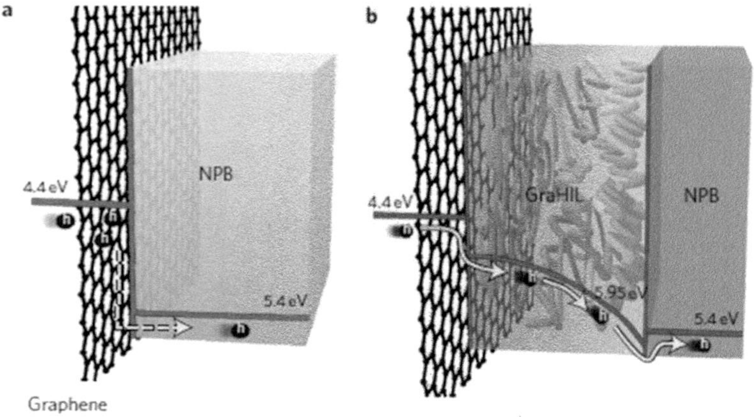

Figure 9.2 Schematic illustrations of hole injection from graphene in an OLED device. (a) Hole injection process from a graphene anode to a conventional HTL (NBP). (b) Hole injection process from a graphene anode via a self-organized HIL to the HTL (NBP). Reprinted by permission from Macmillan Publishers Ltd: *Nature Photonics*, Ref. 13, Copyright 2012.

HIL they used was composed of poly(3,4-ethylenedioxythiophene) doped with poly(styrenesulphonate) (PEDOT:PSS) and a kind of perfluorinated ionomers. This HIL provided a WF gradient through the layers and thus largely improved the hole-injection efficiency from graphene to the overlying organic layer [5]. Based on these developments, the current efficiencies and the luminous efficiencies of OLEDs are all improved. Current efficiencies were 30.2 cd/A and 98.1 cd/A and luminous efficiencies were 37.2 lm/W and 102.7 lm/W in flexible fluorescent and phosphorescent OLEDs with four-layered graphene, respectively. These remarkable device efficiencies signal a great potential for OLEDs with graphene anodes [5].

Graphene can also be used as a transparent cathode electrode of LED [18]. Hwang et al. had prepared an N-doped reduced graphene electrode as an alternative to fluorine-doped tin oxide (FTO) for high-performance inverted polymer LED (iPLED) [15]. The device fabrication process is shown in Fig. 9.3. This N-doped reduced graphene had a relatively low WF (~4.25 eV) compared to other conventional transparent electrodes (4.5–5.2 eV), which offered a valuable transparent cathode for inverted-type optoelectronic

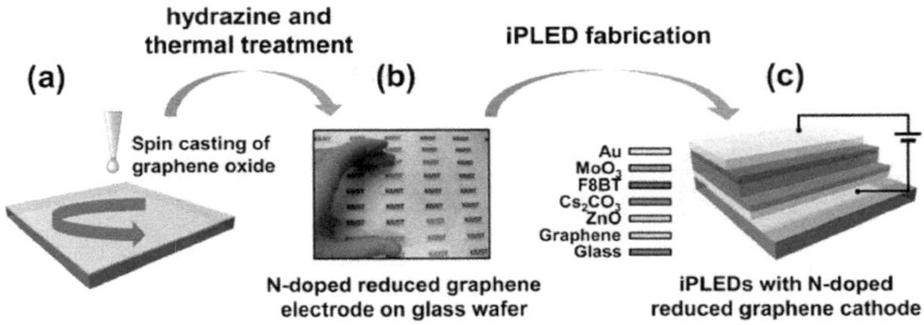

Figure 9.3 Schematic description of preparation and optoelectronic transparent electrode utilization of N-doped reduced graphene film. Reprinted with permission from Ref. 15, Copyright 2012, American Chemical Society.

devices with a minimized electron injection barrier [15]. In contrast, devices with thermally reduced graphene cathodes without N-doping illustrated an extremely larger turn-on voltage and lower luminance than other devices, which is resulted from the large sheet resistance and significant injection barrier of the reduced graphene caused by the mismatch in energy level with ZnO electron transport layer [15].

9.2.2 Improvement in CVD Growth of Graphene

Graphene synthesized via CVD requires high processing temperature, which limits the number of substrates that can be used for the graphene growth. In addition, ohmic contact is also critical for full performance of the devices. Therefore, to integrate graphene directly into optoelectronic devices, it is essential to reduce the growth temperature and to form ohmic contact between graphene and the other components.

In order to improve the performance of graphene-based LED, Kim et al. used plasma-assisted metal-catalyst-free synthesis for graphene/GaN-based LEDs [11]. The graphene growth on the metal-catalyst-free substrate was through plasma-enhanced CVD (PECVD), as shown in Fig. 9.4 [11]. During the plasma-assisted process, methane is effectively dissociated into various species, such as CH_x, C_2H_y, H, and H_2. By changing the plasma power, the dissociation rate of methane can be tuned. When the plasma power is at

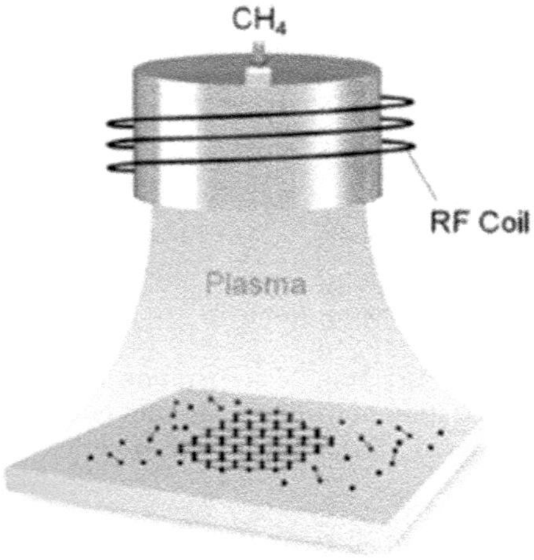

Figure 9.4 Schematic illustration of the graphene growth process in PECVD system. Reprinted with permission from Ref. 11, Copyright 2014, American Chemical Society.

50 W, the dissociation rate of methane powder is around 34%, which is extremely higher than that of methane (~0.0002%) in a thermal CVD process. The growth temperature of graphene was successfully reduced to 600°, which is low enough to prevent thermal degradation of active layers in LEDs [11].

Moreover, when polycrystalline graphene was directly integrated into GaN-based LEDs through PECVD, in situ formation of ohmic contact was found between graphene and p-GaN. Figure 9.5 illustrates the current–voltage (I–V) characteristics of contacts between DG and p-GaN and transferred graphene (TG) and p-GaN contacts. TG/p-GaN shows a strong nonlinear behavior, while DG/p-GaN demonstrates ohmic behavior even in the low-voltage range. The behavior of TG/p-GaN could be attributed to factors such as adhesion, the Schottky barrier between the two layers, and the evenness of the graphene surface in contact with GaN. The behavior of DG/p-GaN could be caused by the ohmic contact between graphene and p-GaN. A possible way to establish ohmic contact is through increased carrier concentration, which can reduce the width of the tunneling barrier and then enhance carrier

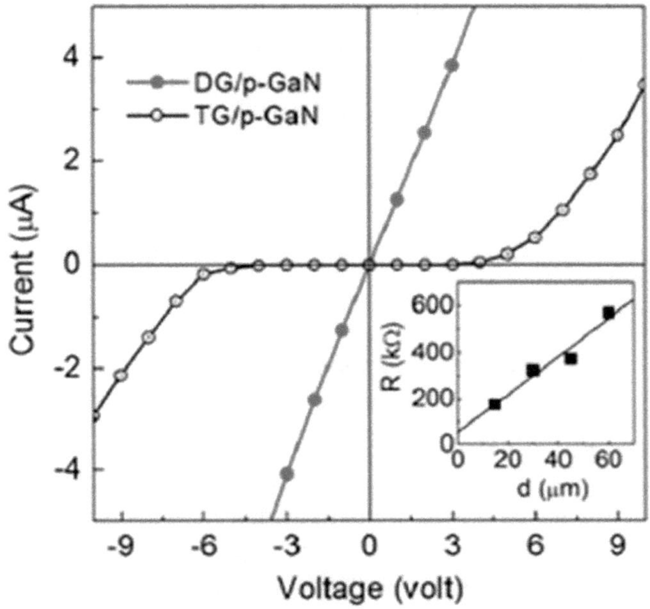

Figure 9.5 I–V characteristics of DG/p-GaN and TG/p-GaN contacts. Reprinted with permission from Ref. 11, Copyright 2014, American Chemical Society.

tunneling through the barrier [2]. In conclusion, PECVD provides a possible method to make high-performance, large-scale, and low-cost graphene-based LEDs.

9.3 Combination of Graphene Electrodes with Other Materials

9.3.1 Integration of Graphene Electrodes with Nanoclusters

In order to improve the performance of graphene-based LEDs, scientists tried to combine graphene with other materials such as nanoclusters. In 2014, Seo et al. combined Ag nanoclusters with graphene film as a transparent and current spreading electrode

(TCSE) [16]. The reason to choose Ag in their experiment is that Ag has a low-resistance ohmic contact with p-GaN and has a similar WF to that of graphene [16]. The structure of the device is shown in Fig. 9.6. In this device, the Ag nanoclusters act as low-ohmic contact sites to the p-GaN as well as current sink sources, while graphene film acts as a transparent conducting network, connecting Ag nanoclusters. It had been found that when using a compound Ag nanocluster–graphene electrode, the performance of the LED had been improved.

9.3.2 Integration of Graphene Electrodes with One-Dimensional Semiconductor Nanostructures

In order to improve the efficiency of LEDs and reduce the cost, nanowires have been researched for three-dimensional LEDs, which can better handle the strain induced by thermal expansion mismatch and an improved extraction efficiency [3]. Besides, by changing the nanowire diameters, it is possible to change the emission color of nanowire LEDs grown in the same run. In addition, single-nanowire LEDs can possibly achieve efficient submicrometer emitters that can be integrated into a functional photonic circuit.

In 2010, Lee et al. combined two-dimensional graphene sheets with air-gap one-dimensional pillar superlattice (PSL) arrays as the top electrode, in which graphene is mounted directly on the PSL to form a freely suspended structure [17] (Fig. 9.7). In this design, graphene sheets could provide efficient current spreading and injection into the active regions of the PSL for light emission, with minimal optical absorption or reflection [17]. Compared with transparent conducting oxides (TCOs), graphene sheets possess additional attractive features such as unusually large mechanical strength and elasticity, and graphene can be used in some circumstances where TCOs are not suitable due to their brittleness.

In PSL array and graphene hybrid LEDs, one problem is that the direct contact of graphene with the inorganic semiconductors generally leads to high injection barriers that degrade performance and frustrate reliable operation. To overcome this limitation,

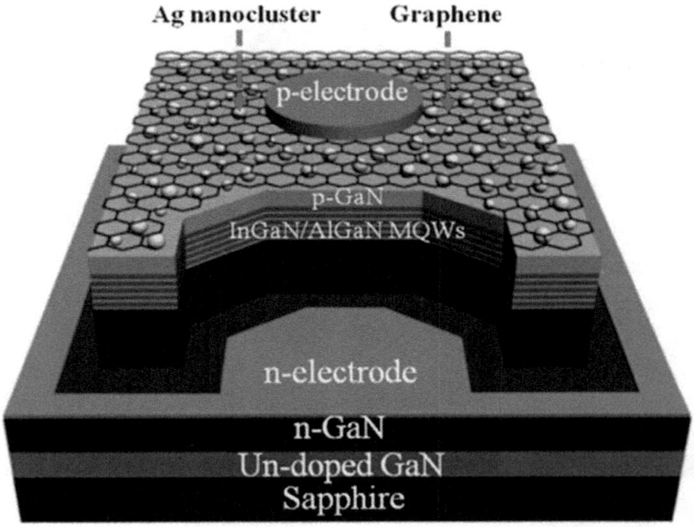

Figure 9.6 Schematic diagram of fabricated InGaN/AlGaN MQWs-based NUV-LED with a CA-G electrode adopted for transparent and current-spreading electrode. Reproduced with permission from Ref. 16, Copyright 2014, IOP Publishing. All rights reserved.

graphene is exploited as a deposition site for metal atoms. In this case, graphene can provide diffusion pathways to the pillar tips, where interfacial reactions lead to improved contact properties for appropriate metals [17].

9.3.3 Integration of Graphene Electrodes with SWCNTs

Huang et al. later combined single-walled carbon nanotubes (SWCNTs) with chemically reduced graphene, in order to modulate the WF of the electrode to form an ohmic contact with the active layer, which can improve charge injection and, therefore, enhance the performance of the device [20]. To make this device, they mixed dry powders of graphene oxide and SWCNTs directly in anhydrous hydrazine and doped these carbon composite films with alkali carbonates to tune the WF of these carbon composite films [20]. After thermal annealing, the alkali carbonates decomposed into alkali

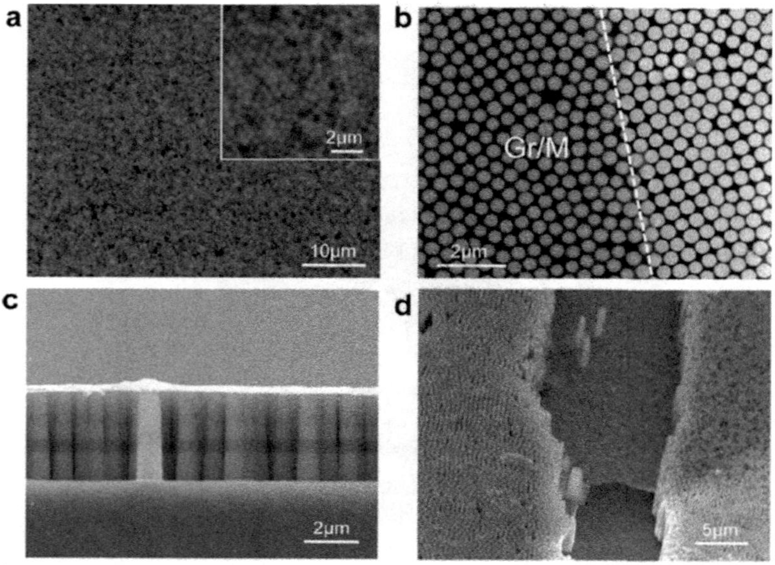

Figure 9.7 Optical images (a) and SEM images (b–d) of the PSL-Gr/M architectures. Reprinted with permission from Ref. 17, Copyright 2010, American Chemical Society.

oxides with low WF, which covered the carbon-based materials [20]. These alkaline-doped carbon-based hybrid electrodes made a new strategy to tune the WF of the device, which boosted the development of flexible and ITO-free LED devices [20].

9.3.4 Integration of Graphene Electrodes with Nanowires

Recently, the core–shell geometry of nanowire has drawn more attention in GaN-based LEDs because this kind of structure allows to increase the active device area leading to droop reduction and to suppress the quantum confined Stark effect in InGaN/GaN quantum walls grown on lateral m-plane facets [19]. The first core–shell InGaN/GaN LED was built from disordered nanowires grown along the m-axis using catalyst-assisted metal organic vapor phase epitaxy (MOVPE) [19]. In 2014, Tchernycheva et al. fabricated single-nanowire InGaN/GaN LEDs with a transparent graphene contact for

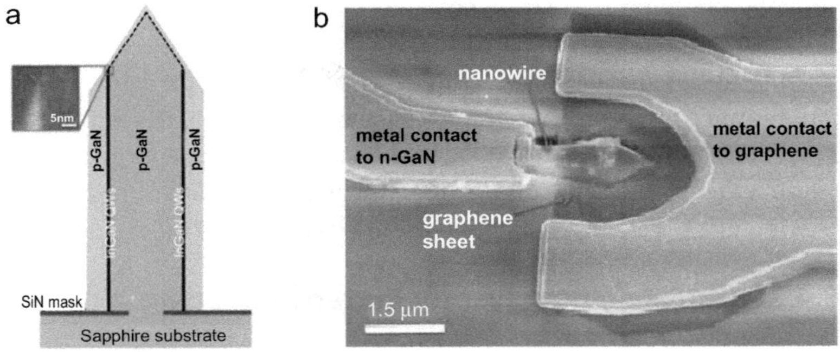

Figure 9.8 (a) Schematic diagram of the nanowire structure. (b) SEM image of the single-wire graphene-contacted LED. Reprinted with permission from Ref. 19, Copyright 2014, American Chemical Society.

hole injection. The structures of core–shell InGaN/GaN nanowires and their device are shown in Fig. 9.8.

9.4 Other Potential Applications of Graphene in LEDs

9.4.1 Photoluminescence from Graphene Oxide

Graphene oxide with blue or red emissions was obtained in 2009 [20]. The photoluminescence of graphene oxide can originate from the recombination of electron–hole pairs in sp^2 clusters (1 nm). Generally, carbon materials contain a mixture of sp^2 and sp^3 bonding. The optical properties of disordered carbon thin films containing a mixture of sp^2 and sp^3 carbon have been widely investigated. The opto-electronic properties of these materials are determined by the π states of the sp^2 sites [21]. The π and π^*, electronic levels of the sp^2 clusters lie within the bandgap of the σ and σ, states of the sp^3 matrix and are strongly localized [21]. The photoluminescence in such carbon systems is a consequence of geminate recombination of localized electron–hole pairs in the sp^2 clusters, which essentially behave as the luminescence centers or chromophores [21]. Since the bandgap depends on the size, shape, and fraction of the sp^2 domains, tunable photoluminescence emission can be achieved by controlling the nature of the sp^2 sites [21].

Due to the special photoluminescence of graphene oxide, it has a potential to be used as luminescence source of LED. Recently, the most common methods to make luminescent graphene or graphene oxide with blue or red emissions are through oxidation or reduction treatments, for example, oxygen plasma treatment [20].

9.4.2 Graphene Interlayer for Current Spreading Enhancement

Due to current spreading layers (CSLs) used in LEDs, the performance of LEDs has significantly improved. Without a CSL, the current flow will converge at the metal pad in GaN-based LEDs. This phenomenon is called current crowding [22]. When a CSL is used, current crowding will be solved due to the lower sheet resistance and proper WF [22]. Recently, graphene has been used as CSLs in GaN-based LEDs [23, 24]. However, compared with other CSL materials such as thin metal layers and ITO, high sheet resistance and low WF of pristine graphene cause a high turn-on voltage with insufficient current spreading [25]. In order to solve this problem, instead of directly using graphene as a CSL, Min et al. put a graphene interlayer between ITO and p-type GaN (p-GaN) to improve current spreading [22]. There are several reasons contributed to this improvement. First, pristine graphene produces more than 95% transmittance in the visible range, which leads to no degradation in transparency as an interlayer. Second, the high mobility of pristine graphene helps carrier transport in the device. Third, the two-dimensional structure of graphene makes no change in the total thickness of the layers [22].

9.5 Conclusion

The excellent physical and chemical properties have made graphene an appealing material for LED technology. Combining graphene with other materials can further eliminate the shortcomings of graphene and lead to high-performance graphene-based LEDs.

References

1. Sze, S. M., and Ng, K. K. (2006). *Physics of Semiconductor Devices*, Wiley.
2. Chang, M.-H., Das, D., Varde, P. V., and Pecht, M. (2012). Light emitting diodes reliability review, *Microelectronics Reliability*, **52**, pp. 762–782.
3. Krames, M. R., Shchekin, O. B., Mueller-Mach, R., Mueller, G. O., Ling, Z., Harbers, G., and Craford, M. G. (2007). Status and future of high-power light-emitting diodes for solid-state lighting, *Journal of Display Technology*, **3**, pp. 160–175.
4. Steigerwald, D. A., Bhat, J. C., Collins, D., Fletcher, R. M., Holcomb, M. O., Ludowise, M. J., Martin, P. S., and Rudaz, S. L. (2002). Illumination with solid state lighting technology, *Selected Topics in Quantum Electronics, Journal of IEEE*, **8**, pp. 310–320.
5. Schubert, E. F., Jong Kyu, K., Hong, L., and Xi, J. Q. (2006). Solid-state lighting: A benevolent technology, *Reports on Progress in Physics*, **69**, pp. 3069.
6. Aoyama, Y., and Yachi, T. (2008). An LED module array system designed for streetlight use, *Energy 2030 Conference*, 2008. ENERGY 2008. IEEE, pp. 1–5.
7. Steranka, F. M., Bhat, J., Collins, D., Cook, L., Craford, M. G., Fletcher, R., Gardner, N., Grillot, P., Goetz, W., Keuper, M., Khare, R., Kim, A., Krames, M., Harbers, G., Ludowise, M., Martin, P. S., Misra, M., Mueller, G., Mueller-Mach, R., Rudaz, S., Shen, Y. C., Steigerwald, D., Stockman, S., Subramanya, S., Trottier, T., and Wierer, J. J. (2002). High power LEDs: Technology status and market applications, *Physica Status Solidi (a)*, **194**, pp. 380–388.
8. Castro Neto, A. H., Guinea, F., Peres, N. M. R., Novoselov, K. S., and Geim, A. K. (2009). The electronic properties of graphene, *Reviews of Modern Physics*, **81**, pp. 109–162.
9. Geim, A. K. (2009). Graphene: Status and prospects, *Science*, **324**, pp. 1530–1534.
10. Wu, J., Agrawal, M., Becerril, H. A., Bao, Z., Liu, Z., Chen, Y., and Peumans, P. (2010). Organic light-emitting diodes on solution-processed graphene transparent electrodes, *ACS Nano*, **4**, pp. 43–48.
11. Kim, Y. S., Joo, K., Jerng, S.-K., Lee, J. H., Moon, D., Kim, J., Yoon, E., and Chun, S.-H. (2014). Direct integration of polycrystalline graphene into light emitting diodes by plasma-assisted metal-catalyst-free synthesis, *ACS Nano*, **8**, pp. 2230–2236.

12. Kumar, A., and Zhou, C. (2010). The race to replace tin-doped indium oxide: Which material will win? *ACS Nano*, **4**, pp. 11–14.

13. Han, T.-H., Lee, Y., Choi, M.-R., Woo, S.-H., Bae, S.-H., Hong, B. H., Ahn, J.-H., and Lee, T.-W. (2012). Extremely efficient flexible organic light-emitting diodes with modified graphene anode, *Nature Photonics*, **6**, pp. 105–110.

14. Sun, T., Wang, Z. L., Shi, Z. J., Ran, G. Z., Xu, W. J., Wang, Z. Y., Li, Y. Z., Dai, L., and Qin, G. G. (2010). Multilayered graphene used as anode of organic light emitting devices, *Applied Physics Letters*, **96**, pp. 133301.

15. Hwang, J. O., Park, J. S., Choi, D. S., Kim, J. Y., Lee, S. H., Lee, K. E., Kim, Y.-H., Song, M. H., Yoo, S., and Kim, S. O. (2012). Workfunction-tunable, N-doped reduced graphene transparent electrodes for high-performance polymer light-emitting diodes, *ACS Nano*, **6**, pp. 159–167.

16. Seo, T. H., Kim, S., Kim, M. J., Kim, H., and Suh, E.-K. (2014). Compound Ag nanocluster-graphene electrodes as transparent and current spreading electrodes for improved light output power in near-ultraviolet light emitting diodes, *Journal of Physics D*, **47**, pp. 215103.

17. Lee, J. M., Choung, J. W., Yi, J., Lee, D. H., Samal, M., Yi, D. K., Lee, C.-H., Yi, G.-C., Paik, U., Rogers, J. A., and Park, W. I. (2010). Vertical pillar-superlattice array and graphene hybrid light emitting diodes, *Nano Letters*, **10**, pp. 2783–2788.

18. Huang, J.-H., Fang, J.-H., Liu, C.-C., and Chu, C.-W. (2011). Effective work function modulation of graphene/carbon nanotube composite films as transparent cathodes for organic optoelectronics, *ACS Nano*, **5**, pp. 6262–6271.

19. Tchernycheva, M., Lavenus, P., Zhang, H., Babichev, A. V., Jacopin, G., Shahmohammadi, M., Julien, F. H., Ciechonski, R., Vescovi, G., and Kryliouk, O. (2014). InGaN/GaN core–shell single nanowire light emitting diodes with graphene-based P-contact, *Nano Letters*, **14**, pp. 2456–2465.

20. Lu, Y., Jiang, Y., Wei, W., Wu, H., Liu, M., Niu, L., and Chen, W. (2012). Novel blue light emitting graphene oxide nanosheets fabricated by surface functionalization, *Journal of Materials Chemistry*, **22**, pp. 2929–2934.

21. Eda, G., Lin, Y.-Y., Mattevi, C., Yamaguchi, H., Chen, H.-A., Chen, I. S., Chen, C.-W., and Chhowalla, M. (2010). Blue photoluminescence from chemically derived graphene oxide, *Advanced Materials*, **22**, pp. 505–509.

22. Min, J.-H., Son, M., Bae, S.-Y., Lee, J.-Y., Yun, J., Maeng, M.-J., Kwon, D.-G., Park, Y., Shim, J.-I., Ham, M.-H., and Lee, D.-S. (2014). Graphene interlayer

for current spreading enhancement by engineering of barrier height in GaN-based light-emitting diodes, *Optics Express*, **22**, pp. A1040–A1050.

23. Kim, B.-J., Lee, C., Jung, Y., Hyeon Baik, K., Mastro, M. A., Hite, J. K., Eddy, C. R., and Kim, J. (2011). Large-area transparent conductive few-layer graphene electrode in GaN-based ultra-violet light-emitting diodes, *Applied Physics Letters*, **99**, pp. 143101.
24. Kim, B.-J., Lee, C., Mastro, M. A., Hite, J. K., Eddy, C. R., Ren, F., Pearton, S. J., and Kim, J. (2012). Buried graphene electrodes on GaN-based ultra-violet light-emitting diodes, *Applied Physics Letters*, **101**, pp. 031108.
25. Kun, X., Chen, X., Jun, D., Yanxu, Z., Weiling, G., Mingming, M., Lei, Z., and Jie, S. (2013). Graphene transparent electrodes grown by rapid chemical vapor deposition with ultrathin indium tin oxide contact layers for GaN light emitting diodes, *Applied Physics Letters*, **102**, pp. 162102.

Chapter 10

Graphene-Based Touchscreens

Shivananju Bannur Nanjunda[a] **and Qiaoliang Bao**[a,b]

[a] *Institute of Functional Nano & Soft Materials (FUNSOM), Soochow University, 199 Ren-Ai Road, Suzhou Industrial Park, Suzhou Jiangsu 215123, China*
[b] *Department of Materials Science and Engineering, Monash University, 22 Alliance Lane, Building 69, Clayton, 3800, Victoria, Australia*
qiaoliang.bao@gmail.com

10.1 Introduction

Touchscreens are visual outputs that can detect the presence and location of a touch within the display area, permitting physical interaction with what is shown on the display itself [1]. Pointing at a thing, or touching it, is one of the most natural ways to select it (Fig. 10.1). Touchscreen is an electronic visual display device that the user can control through simple or multi-touch gestures by touching the screen with a special stylus/pen and or one or more fingers (Fig. 10.1). It is a means of capturing a human's natural pointing instinct and using it as a mode of human–computer communication [1]. Touchscreens are easy to use and require no additional work space and no moving parts such as keyboard and mouse system.

Graphene Photonics, Optoelectronics, and Plasmonics
Edited by Qiaoliang Bao, Hui Ying Hoh, and Yupeng Zhang

ISBN 978-981-4745-60-4 (Hardcover), 978-1-315-19667-1 (eBook)
www.panstanford.com

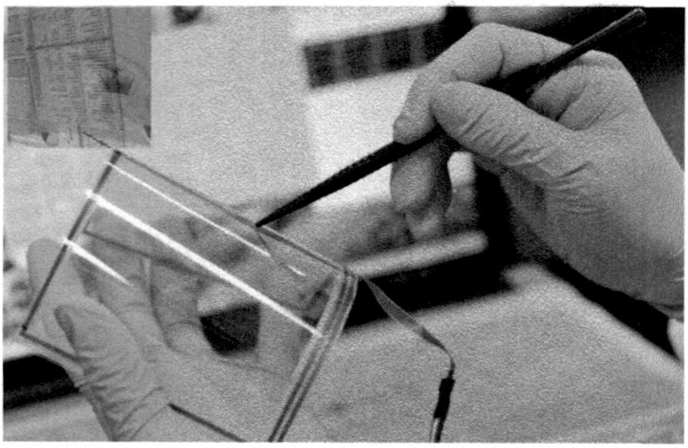

Figure 10.1 Graphene-based touchscreen. Reprinted by permission from Macmillan Publishers Ltd: *Nature Nanotechnology*, Ref. 7, Copyright 2010. Inset image courtesy: http://cdni.wired.co.uk/620x413/s_v/shutterstock_61448245.jpg

Touchscreens are becoming more popular and are currently used in a wide range of applications such as cellular phones, digital cameras, Amazon Kindle, airports, shopping malls, banking, hospitals, amusement parks, and home automation, because they allow quick, intuitive, and accurate interaction by the user with the display content [1].

10.2 Transparent Touchscreen Electrodes

Transparent electrodes are a necessary component in many modern devices such as touchscreens, liquid crystal display (LCD), solar panels, and organic light-emitting diodes (OLED) [2]. Traditionally, this role has been well served by doped metal oxides, the most common of which is indium tin oxide, commonly known as ITO [2]. Recently, advances in nanomaterials research have opened the door for other transparent conductive materials, each with unique properties such as carbon nanotubes (CNTs) and graphene. These emerging nanomaterial alternatives to ITO will find room in the marketplace [2].

The transparent conductive film (TCF) requirements for touchscreens are $R_S \leq 500-2000\ \Omega$ and $T > 90\%$ at 550 nm [3], favorable mechanical properties, high chemical durability, no toxicity, and low production costs [2,3]. The commonly used transparent electrodes in touchscreens are ITO, conducting polymers, CNTs, and graphene.

10.2.1 Indium Tin Oxide

Tin-doped indium oxide (ITO) has been the dominant material used as transparent electrodes in touchscreen in the 20th century [2]. In the 1930s, scientists at corning laboratories were investigating novel transparent conducting oxides in part by accident (US Patent #2118795, filed September 21, 1931) during experimentation with metal oxide insulators. Extensions of this work were advanced in the 1940s (US Patent #2564707) with clear mentions of indium and tin oxide mixtures. ITO has been studied and refined for more than 50 years, and as a result, the material offers many beneficial properties that have made it the material of choice [2, 3].

However, ITO has lots of problems, such as scarcity of supply (which is leading to dramatic price fluctuations), its ceramic nature (which leads to brittleness and easily damaged films), and micro cracks can propagate eventually leading to sharp decreases in the film's electrical conductivity [2]. Cracking of the film during manufacturing, handling, and device assembly leads to significant yield loss even before the device reaches the consumer. Furthermore, corrosion of the ITO layer due to trace amounts of salts or acids in the environment or device adhesives can decrease the cause lifetime [2]. Another potential disadvantage with ITO is its relatively high index of refraction ($n \sim 2.0$), which leads to unwanted reflection when coupled with lower index substrates and materials [2]. Although antireflective coating stacks have been developed to elevate this problem, but these can be quite expensive. The high cost associated with ITO materials and deposition process is perhaps the greatest threat to continue with ITO material [2].

10.2.2 Conducting Polymers

Polyacetylene was the first documented polymer to demonstrate high conductivity with the addition of p or n dopants [2]. Conductive polymers gained prominence as a viable technology in the 1980s when a broad range of commonly available polymers (polypyrrole, polyaniline, polythiophene, etc.) were found to exhibit significant electrical conductivities via simple chemical-doping mechanisms [2].

A problem with conductive polymers is the instability of the doped state, which can rapidly decrease the electronic conductivity level when exposed to various stresses such as temperature, humidity, UV light, and chemicals [2]. This instability, no doubt, has been one of the main reasons that conductive polymers have not achieved widespread commercial application in touch panels, displays, and other devices, despite lab-scale demonstration of such devices [2].

10.2.3 Carbon Nanotubes

The first experimental observations of CNTs were made in the early 1990s [4], and fundamental transport properties were studied during the subsequent decade [5]. Individual CNTs were found to have mobilities in excess of 100,000 cm^2/Vs, current-carrying capacity of 10^9 A/cm^2, and ON/OFF current ratios higher than 10^5 [2]. However, despite their amazing electrical, optical, and mechanical properties, CNTs have been slow to find a commercial use. One of the major obstacles toward implementing CNTs is the inability to obtain bulk quantities of material with suitable purity [2]. CNTs, as produced, typically contain a mixture of various materials, including catalyst particles, catalyst support, amorphous or non-tubular carbon, and a mixture of nanotubes of various lengths, diameters, tube number, and chiralities. The inability to produce pure CNTs in a large scale with low cost remains one of the major hurdles to continue with CNTs [2].

10.2.4 Graphene

Graphene is an amazing material with outstanding electrical, optical, mechanical, and chemical properties making it attractive for applications in optoelectronic devices such as transparent electrodes in touchscreens, displays, and photovoltaic modules [6,7].

A pure single-layer graphene is a 0 eV bandgap semiconductor with a thickness of 0.34 nm. Electrons in a single graphene delocalize over the sheet, which leads to a very high Fermi velocity $v_F = 10^6$ m/s and a high in-plane conductivity. Firsov et al. found that a graphene film exhibits a strong ambipolar electric field effect showing electron/hole concentrations up to $10^{13}/cm^2$ and room-temperature mobilities of $\sim 10^4$ cm^2/Vs. Peumans et al. predicted that the sheet resistance of graphene will vary with the number of layers as $R_s \sim 62.4/N\Omega/sq$ for highly doped graphene, where N is the number of layers and the transmittance will vary as $T \sim 100-2.3N$ (%) [2, 6, 7].

In this chapter, we will focus on the aspects of graphene as a transparent electrode for touchscreen applications.

10.2.5 Working Principle of Graphene Touchscreen

The most common touchscreen device formats are resistive and capacitive, both of which consist of various layers such as antireflective coating, transparent electrodes, resistive or capacitive sensing circuit, glass/polymer substrate, and LCD (Figs. 10.2a,b) [6–10].

Figure 10.2c gives the overall function of how iPhone interprets touch signal. After collecting touch signal data from the screen, the iPhone processor must interpret the raw touch data into a gesture. This instantaneous process is broken into several steps: (1) The screen interprets a touch on the screen as electrical signals. These electrical signals are sent to the iPhone processor. (2) The iPhone processor software analyses the size, shape, and location of the touch. If you swipe your finger along the screen, the software determines the start and end points of the touch. (3) The processor uses "gesture-interpretation" software to determine the type of touch on the screen (i.e., swipe, scroll, flick, etc.) (4). The processor relays the data to the program that you are using [10].

10.3 Resistive Touchscreens

Resistive touchscreens work on the basis of pressure applied to the screen (Fig. 10.3). When the screen is pressed, the outer layer is

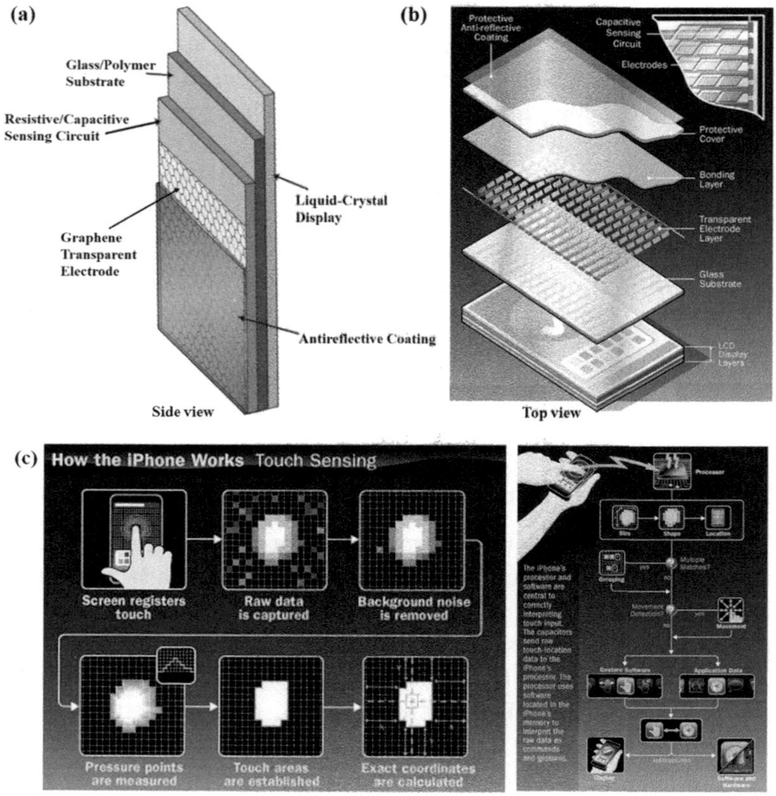

Figure 10.2 Schematic diagram of graphene resistive/capacitive touch-screen: (a) side view, (b) top view, (c) How the iPhone touchscreen works. Figure (a) reprinted by permission from Macmillan Publishers Ltd: *Nature Photonics*, Ref. 6, Copyright 2010. Figures (b) and (c) reprinted from Ref. 10.

pushed onto the next layer; the technology senses that pressure is being applied and registers input [1, 2, 16].

Usually a resistive graphene touchscreen consists of at least three layers: A flexible membrane made from polyester or polyethylene terephthalate (PET) film is suspended over a rigid substrate made from glass/polymer substrate [2, 7]. Both surfaces are coated with a TCF like graphene. The transparent conductive graphene layers are kept apart by periodic insulating spacer dots, which reinforce an air gap between the two graphene layers (Figs. 10.3a,b).

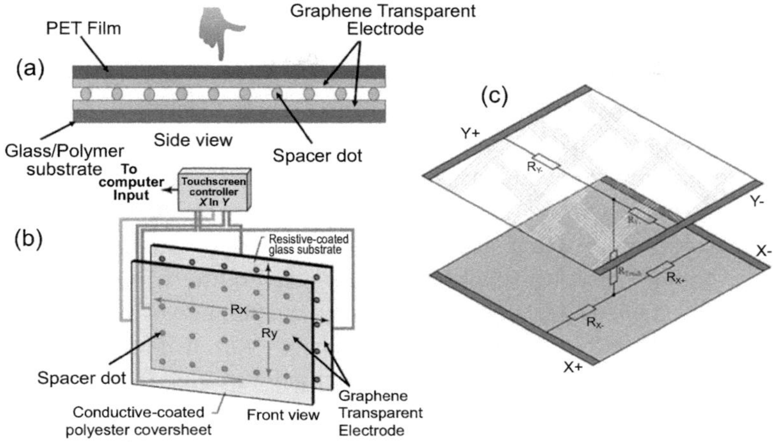

Figure 10.3 Schematic diagram of basic four-wire resistive touch panel device structure and functional layers: (a) side view, (b) front view, (c) the transparent conductor arrangement for a four-wire analog resistive touch panel. Figure (c) reprinted from Ref. 2 with permission from John Wiley and Sons.

Pushing down on the outer membrane "short circuits" the two conductive graphene layers at an intersection point, thereby defining the touch-placement location [2, 7]. The bus bars in the top sheet and substrate are perpendicular to each other. In this way, there will be no electrical connection unless pressure is applied to the top sheet (PET film). The bus bars are connected to the touchscreen controller through a four-wire flex cable. The four wires are referred to as X+ (left), X− (right), Y+ (top), and Y− (bottom) [2]. The point of contact "divides" each layer in a series resistor network with two resistors and a connecting resistor between the two layers (Fig. 10.3c). By measuring the voltage at this point, the user gets information about the position of the contact point orthogonal to the voltage gradient. To get a complete set of coordinates, the voltage gradient must be applied once in vertical and then in horizontal directions: First a supply voltage must be applied to one layer, and a measurement of the voltage across the other layer is performed. Next the supply is instead connected to the other layer, and the opposite layer voltage is measured. In the standby mode, one of the lines are connected to a level-triggered interrupt in order to detect touch activity [2, 10].

An advantage of the four-wire touchscreens is that it is possible to determine the touch pressure by measuring the contact resistance (R_{Touch}) between the two graphene layers. R_{Touch} decreases as the touch pressure increases. This characteristic can be useful in applications in which it is required to detect not only where the pressure is applied, but also the type of pressure (area and force).

Although resistive touchscreens are often quite responsive and cost effective, especially in many new smartphones hitting the market, capacitive touchscreens usually provide a more pleasant user experience. Actions such as swiping through contact lists, zooming in and out of web pages and maps, typing e-mails and text messages, and scrolling through photos are best suited to capacitive touchscreens; unlike resistive screens, you can swipe across them gently and still get a response.

10.4 Capactive Touchscreens

Capacitive touchscreens work by sensing the conductive properties of an object, usually the skin on our fingertip (Fig. 10.4), and do not rely on pressure [12,13]. The simplest form of a graphene capacitor touchscreen consists of two conductors, e.g., two graphene transparent electrodes placed in X and Y layers, separated by an insulator such as spacer dots, which act as an air gap (Fig. 10.4a). The following formula shows the parameters that influence capacitance: $C = \varepsilon\ (A/d)$ and $\varepsilon = \varepsilon_0{}^{*}\ \varepsilon_r$, where C is the capacitance, ε_r is the relative permittivity, also called dielectric constant of the insulating spacer dot materials between the graphene plates, ε_0 is the permittivity of free space (8.854×10^{-12} F/m), A is the area of the graphene plates, and d is the distance between the graphene plates.

The graphene electrode of a touch sensor represents one plate of such a capacitor. The corresponding second plate is represented by the environment of the sensor electrode (to form a parasitic capacitor C_x) and another conductive object, like a human finger (to form touch capacitor C_g) (Fig. 10.4b). This capacitor, i.e.,

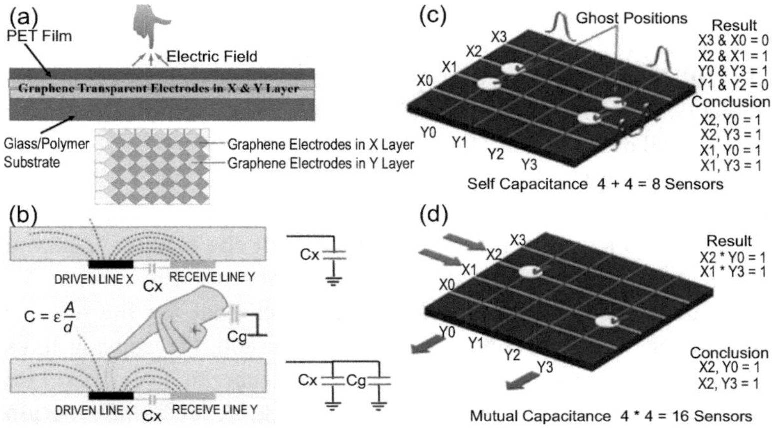

Figure 10.4 Schematic of basic capacitive touch panel device structure and functional layers: (a) device structure, (b) working principle, (c) self-capacitance, and (d) mutual capacitance. Reprinted from Ref. 11.

the sensor electrode, is connected to a measurement circuit. The capacitance of the sensor pad is measured periodically. If a conductive object (finger) approaches or touches the graphene electrode, the measured capacitance will increase. This change is detected by the measurement circuit and converted into a trigger signal [10, 12].

Device manufacturers have two basic options for arranging and measuring capacitance changes in a touchscreen: self capacitance and mutual capacitance. Most early capacitive touchscreens relied on self capacitance (Fig. 10.4c), which measures an entire row or column of electrodes for capacitive change. This approach is fine for one-touch or simple two-touch interactions. But it presents serious limitations for more advanced applications, because it introduces positional ambiguity when the user touches down in two places. Effectively, the system detects touches at two x-coordinates and two y-coordinates but has no way to know which x goes with which y. This leads to "ghost" positions when interpreting the touch points, reducing accuracy and performance (Fig. 10.4c).

Alternatively, mutual-capacitance touchscreens use transmit and receive electrodes arranged as an orthogonal matrix, allowing them to measure the point where a row and column of electrodes intersect (Fig. 10.4d) [12]. In this way, they detect each touch as a specific pair of (x, y) coordinates. For example, a mutual-capacitance system will detect two touches as (x_1, y_3) and (x_2, y_0), whereas a self-capacitance system will detect simply (x_1, x_2, y_0, y_3).

The iPhone uses both mutual-capacitance and self-capacitance technologies to achieve multi-touch gestures. Mutual capacitance requires two layers of material: one layer for the driving lines and one layer for the sensing lines. Driving lines provide current for the touchscreen. Sensing lines detect the current for the electric node layer. When the finger touches the screen, the sensing lines determine the location of where the touch occurred. Every point on the grid generates its own signal. From the grid, touch information is relayed back to the iPhone's processor. The processor is integral in how the system interprets a touch on the screen. The sensing lines and driving lines create a coordinate grid for the iPhone CPU. This allows the phone to locate the specific area where the touch occurred. Because iPhone relies on a capacitive touchscreen, the screen operates only when a bare finger touches the screen. A gloved hand or stylus will not activate a capacitive touchscreen. The combination of mutual- and self-capacitance technologies allows the iPhone processor to register multi-touch functions [10].

10.5 Graphene-Based Touchscreens

Figure 10.5 shows a screen-printing process used to fabricate four-wire touchscreen panels based on graphene/PET transparent conducting films [7]. After printing electrodes and dot spacers, the upper and lower panels are carefully assembled and connected to a controller installed in a laptop computer (Figs. 10.5a–c), which shows extraordinary flexibility.

Most resistive touch panels are designed to operate with a minimum threshold activation force. The activation force is a function of several film factors. Since the emerging conductors typically have a polymer interspersed at the conductive layer surface, it can add additional interfacial resistance [7]. As the two

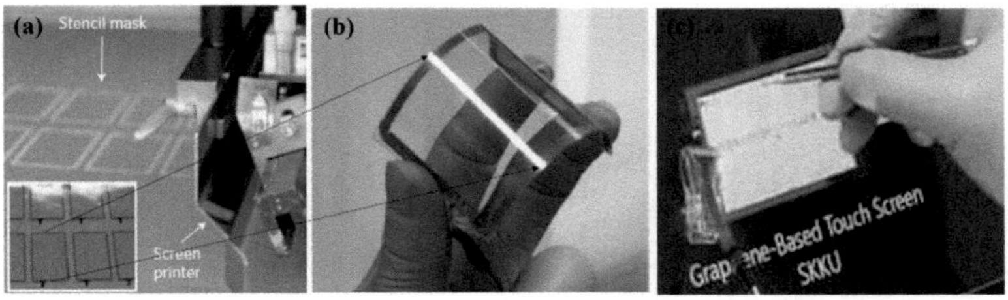

Figure 10.5 Graphene-based touchscreen device. (a) Screen-printing process of silver paste electrodes on graphene/PET film. The inset shows 3.1-inch graphene/PET panels patterned with silver electrodes before assembly. (b) An assembled graphene/PET touch panel showing outstanding flexibility. (c) A graphene-based touchscreen panel connected to a computer with control software. Reprinted by permission from Macmillan Publishers Ltd: *Nature Nanotechnology*, Ref. 7, Copyright 2010.

opposing graphene layers are brought into contact by an input touch, current flows between the two conductive films allowing the controller to identify where the touch position is located and then feed that to the computer software processing the event [7].

10.5.1 Comparison of Graphene and ITO-Based Touchscreen Devices

The electromechanical properties of graphene/PET touchscreen panels were tested (Fig. 10.6).

Unlike an ITO-based touch panel, which easily breaks under 2–3% strain, the graphene-based panel resists up to 6% strain; this is limited not by the graphene itself, but by the printed silver electrodes [7, 14].

Figure 10.6 demonstrates the superior mechanical properties of a graphene film in the context of actual device performance, showing that the resistance between the top and bottom electrodes upon repeated actuation remains unaffected when using a graphene film top electrode but degrades sharply when using an ITO top electrode [7]. These results indicate that graphene films can provide larger viewable area as graphene is not subject to this mechanical vulnerability, even at the extreme device edge points.

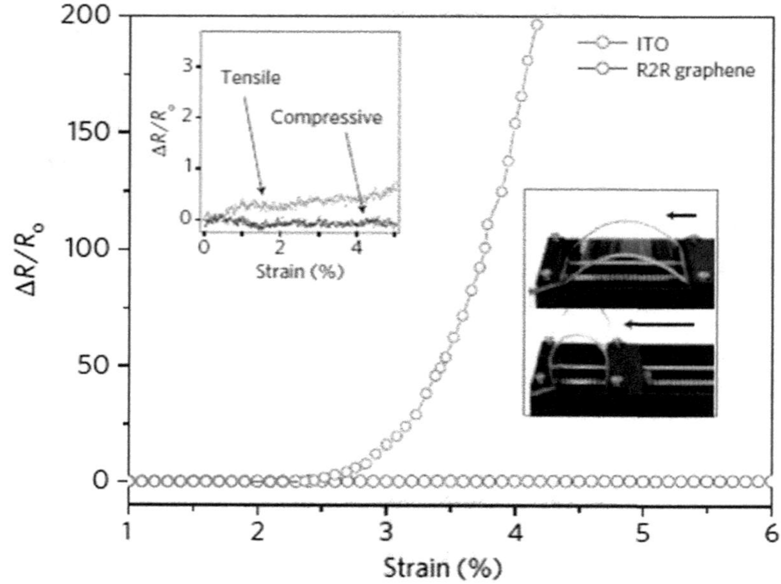

Figure 10.6 Electromechanical properties of graphene-based touchscreen devices compared with ITO/PET electrodes under tensile strain. The inset shows the resistance change with compressive and tensile strain applied to the upper and lower graphene/PET panels, respectively. Reprinted by permission from Macmillan Publishers Ltd: *Nature Nanotechnology*, Ref. 7, Copyright 2010.

10.5.2 Current and Future Trends of Graphene-Based Touchscreens

Graphene-based touchscreens are currently in production line and reaching end users [7]. Patents are filed based on graphene touchscreen [11]. We expect that the market for flexible graphene touchscreens will become larger than that for non-flexible ITO touchscreens in about 10 years [15]. Although there are many candidates for ITO replacement, including metal meshes and silver nanowires whose sheet resistances are as low as ~10 Ω/sq at ~90% transmittance, none of them are as flexible as graphene. For example, the conducting properties of graphene can remain stable up to ~11% extension, and its initial sheet resistance persists for ~300,000 cycles at 3% bending strain, whereas a silver nanowire breaks easily at strains of less than 1% [15]. Therefore, we should expect to see graphene in touchscreens within the next few years.

References

1. Pickering, J. (1986). Touch-sensitive screens: The technologies and their application, *International Journal of Man-Machine Studies*, **25**, pp. 249–269.
2. Hecht, D. S., Hu, L., and Irvin, G. (2011). Emerging transparent electrodes based on thin films of carbon nanotubes, graphene, and metallic nanostructures, *Advanced Materials*, **23**, pp. 1482–513.
3. Sheng, S., Fang, G., Li, C., Xu, S., and Zhao, X. (2006). P-type transparent conducting oxides, *Physica Status Solidi (a)*, **203**, pp. 1891–1900.
4. Iijima, S., and Ichihashi, T. (1993). Single-shell carbon nanotubes of 1-nm diameter, Nature, **363**, pp. 603–605.
5. Wilder, J. W., Venema, L. C., Rinzler, A. G., Smalley, R. E., and Dekker, C. (1998). Electronic structure of atomically resolved carbon nanotubes, *Nature*, **391**, pp. 59–62.
6. Bonaccorso, F., Sun, Z., Hasan, T., and Ferrari, A. C. (2010). Graphene photonics and optoelectronics, *Nature Photonics*, **4**, pp. 611–622.
7. Bae, S., Kim, H., Lee, Y., Xu, X., Park, J. S., Zheng, Y., Balakrishnan, J., Lei, T., Kim, H. R., Song, Y. I., Kim, Y. J., Kim, K. S., Ozyilmaz, B., Ahn, J. H., Hong, B. H., and Iijima, S. (2010). Roll-to-roll production of 30-inch graphene films for transparent electrodes, *Nature Nanotechnology*, **5**, pp. 574–578.
8. Tian, H., Yang, Y., Xie, D., Ren, T. L., Shu, Y., Zhou, C. J., Sun, H., Liu, X., and Zhang, C. H. (2013). A novel flexible capacitive touch pad based on graphene oxide film, *Nanoscale*, **5**, pp. 890–894.
9. Lee, X., Yang, T., Li, X., Zhang, R., Zhu, M., Zhang, H., Xie, D., Wei, J., Zhong, M., Wang, K., Wu, D., Li, Z., and Zhu, H. (2013). Flexible graphene woven fabrics for touch sensing, *Applied Physics Letters*, **102**, pp. 163117.
10. (a) Wilson, T. V., Chandler, N., Fenlon, W., and Johnson, B. How the iPhone works. Available at, http://electronics.howstuffworks.com/iphone.htm. (b) Lorex-Technology. How touch screen monitors works. Available at, https://www.lorextechnology.com/support/self-serve/how-touch-screen-monitorsworks/3100030.
11. http://www.edn.com/design/consumer/4363795/A-magic-touch-The-concept-s-sound-but-implementation-options-abound. (Courtesy Atmel)

12. Akhtar, H., and Kakarala, R. (2014). A methodology for evaluating accuracy of capacitive touch sensing grid patterns, *Journal of Display Technology*, **10**, pp. 672–682.
13. Harley, J. A. *Capacitive Touchscreen or Touchpad for Finger or Stylus*. 2008, Google Patents.
14. Cairns, D. R., Witte II, R. P., Sparacin, D. K., Sachsman, S. M., Paine, D. C., Crawford, G. P., and Newton, R. (2000). Strain-dependent electrical resistance of tin-doped indium oxide on polymer substrates, *Applied Physics Letters*, **76**, pp. 1425–1427.
15. Ahn, J. H., and Hong, B. H. (2014). Graphene for displays that bend, *Nature Nanotechnology*, **9**, pp. 737–738.
16. Lee, K., Hong, W.-K., Choi, M., and Nam, S. (2013). *Touch Display Apparatus Sensing Touch Force*. Google Patents.

Chapter 11

Perspective

Hui Ying Hoh[a] **and Qiaoliang Bao**[a, b]
[a]*Department of Materials Science and Engineering, Monash University, 22 Alliance Lane, Building 69, Clayton, 3800, Victoria, Australia*
[b]*Institute of Functional Nano & Soft Materials (FUNSOM), Soochow University, 199 Ren-Ai Road, Suzhou Industrial Park, Suzhou Jiangsu 215123, China*
qiaoliang.bao@gmail.com

We have brought the readers through a brief yet exciting history of graphene, discussed its optical properties, and highlighted probable applications. Many of these applications are familiar to the common end users, such as solar cells, light-emitting devices, and touchscreens. Developments in processing technology and techniques have made graphene a viable replacement for conventional materials. For example, in photodetectors and solar cells, graphene is highly touted as a candidate for transparent conductors over indium tin oxide (ITO) due to an economical advantage and high mechanical flexibility. Graphene applications are propelled by progress in the production of graphene with suitable properties and manageable cost for the targeted purpose. Nevertheless, the switch to, or even simply integration of, a new material and hence technology is an extensive and costly process. The question to ask next is: Is graphene worth it?

Graphene Photonics, Optoelectronics, and Plasmonics
Edited by Qiaoliang Bao, Hui Ying Hoh, and Yupeng Zhang

ISBN 978-981-4745-60-4 (Hardcover), 978-1-315-19667-1 (eBook)
www.panstanford.com

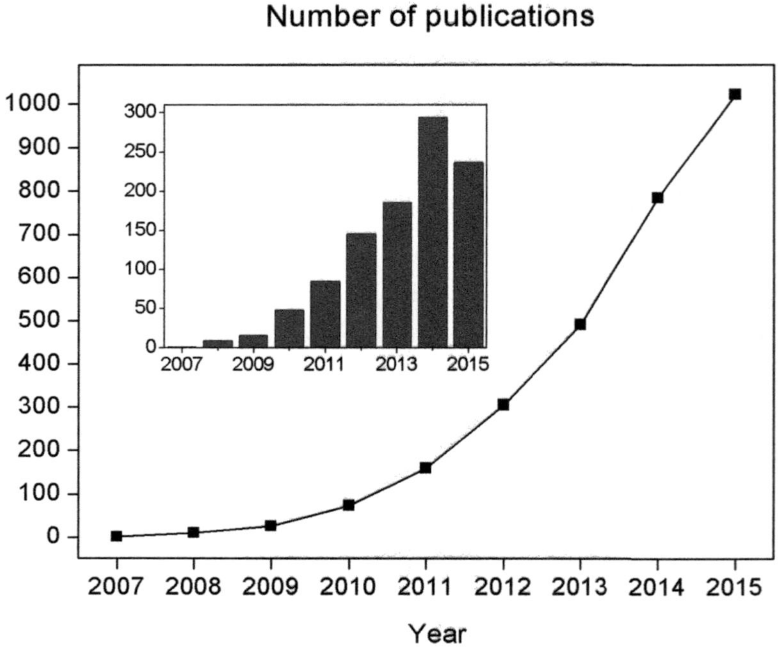

Figure 11.1 Number of publications with the words "graphene," "photonics," "optoelectronics," and "plasmonics" occurring in the titles, abstracts, or keywords from 2007 to date (cumulative). The inset shows the number published each year. The data are obtained by searching for those words in the Scopus database. Note that the data for 2015 are until October 2015.

As we see in Fig. 11.1, the number of publications with the words "graphene," "photonics," "optoelectronics," and "plasmonics" occurring in the titles, abstracts, or keywords has been increasing steadily since 2007. This trend is likely to continue for the next decade or so. The field of graphene photonics, optoelectronics, and plasmonics, despite abundant research efforts in recent years, is still an emerging one. To compare, silicon technology is at most 60 years old, considering the first transistor developed in 1954. Kim et al. have examined a few silicon-based semiconductor devices in which graphene may be incorporated [1]. However, several challenges must be overcome in order to take the current scene to the next level.

11.1 Strategies for Overcoming Challenges

Figure 11.2 summarizes the various graphene-based applications we have discussed and a projection of when a functional device prototype could be expected. We expect that applications utilizing graphene with less stringent requirements will achieve implementation earliest, as in the case of touchscreens or OLEDs. In these applications, the issues to be addressed include better control of the contact resistance and reduction of the sheet resistance [2], both of which should be attainable within these couple of years. Optical limiting products are also expected to appear soon since a range of production methods allow different types of graphene for various optical limiting needs.

For photonic applications, namely, lasers, modulators, and polarizers, a cost-effective production method of high-quality graphene needs to be developed, especially in the case of graphene-based lasers. In addition, a highly integrated optical interconnection between graphene and other materials such as optical fibers and silicon is required. In optoelectronic applications such as photodetectors and solar cells, multiple photocurrent generation mechanisms in graphene should be considered for effective exploitation in specific devices. The overall responsivity of the device also has to be improved. Current strategies include doping control, exploring hybrids or heterostructures, nanostructures, e.g., plasmonic nanostructures and novel device configurations. For example, Chen et al. considered a graphene two-dimensional (2D) homojunction in photoelectric devices. By vertically stacking four graphene monolayers with asymmetric metal contacts, the authors fabricated broadband photoelectric devices, which are simple, easy to scale up, and have a large photo-responsive active area [3]. We will discuss other 2D materials in the following section. More examples of photodetectors based on graphene, other 2D materials, and hybrid systems may also be found in the review by Koppens et al. [4].

In the field of graphene plasmonics, Christensen et al. examined the role of Kerr nonlinearities for graphene plasmonics in nanoribbons and found that at high field strengths, the nonlinearities are strongly enhanced near plasmonic resonances due to field

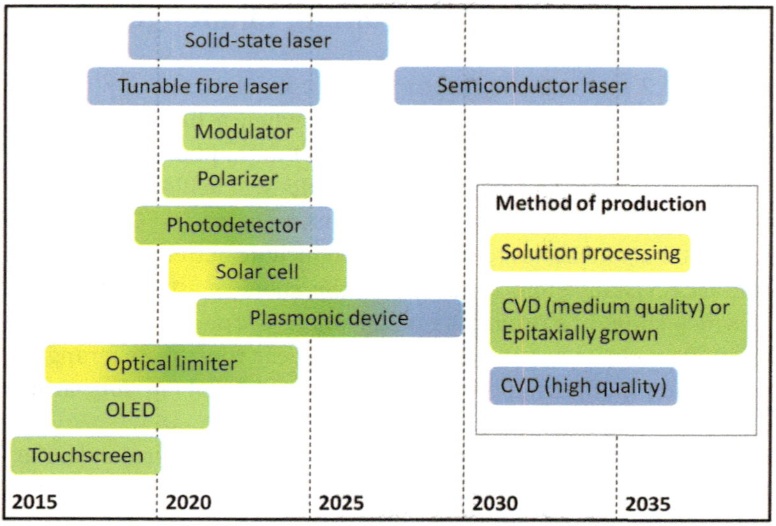

Figure 11.2 Timeline of graphene-based applications. A projection of the timeframe when a functional device prototype could be expected based on device roadmaps and development schedules of industry leaders. Reproduced by permission from Macmillan Publishers Ltd: *Nature*, Ref. 2, Copyright 2012.

enhancement [5]. Jadidi et al. found a new type of plasmon resonance that occurs when graphene is connected to a metal. These new plasmon modes allow one to incorporate a tunable plasmonic channel into a device with electrical contacts, which is a critical step toward realizing practical graphene terahertz optoelectronics [6]. Readers are encouraged to refer to the review by Grigorenko et al. for more information [7].

11.2 Other Two-Dimensional Materials

The rise of graphene has opened doors to a new class of atomically thin or 2D van der Waals materials, for example, boron nitride (BN) [8, 9], transition metal oxides [10], and transition metal dichalcogenides (TMDCs) [11, 12], which formed layered structures weakly held together in varying stacking order and coordination.

Introduction of these 2D materials allows hybrid layers to be prepared from different materials with less concern for lattice mismatch compared to traditional heterostructures. Several reviews on these 2D materials have been published [13–16], and since an in-depth discussion is beyond the scope of this book, we will highlight some significant discoveries here.

One class of material that renders attention is TMDCs, which have been studied for decades and re-visited recently [11, 12]. TMDCs offer promising properties, such as direct bandgaps of around 1–2 eV in molybdenum disulfide (MoS_2) and tungsten diselenide (WSe_2), encouraging light-emitting properties in the near-infrared range [17]. More interestingly, TMDCs demonstrated unique physics due to the lack of an inversion center in the crystal structure [14], leading to valley-contrasting orbital magnetic moment and circular dichroism [18–20]. Polarized LEDs in TMDC monolayer p-n junctions have been reported [21–23]. Strong and long-lived excitons in TMDC monolayers are also attractive for LEDs and related applications. In fact, trions have been observed in doped TMDC monolayers [24,25]. Trions are quasiparticles consisting of two electrons and a hole or two holes and an electron. Since trions are charged, they may be manipulated by electric fields, allowing efficient charge transport and collection of photo-generated current in photodetectors and solar cells [26, 27].

Other 2D materials that come into the limelight recently include topological insulators and layered black phosphorus. For example, we reported a graphene–Bi_2Te_3 photodetector with a much higher photoresponsivity and sensitivity compared to the pure graphene-based devices. Due to the effective photocarrier generation and transfer at the interface between graphene and Bi_2Te_3, the generated photocurrent is enhanced. In the case of layered black phosphorus, with a bandgap of 0.3 eV, it acts as an intermediate between gapless graphene and other 2D materials such as TMDCs [28]. One fascinating property of black phosphorus is its in-plane anisotropy, affecting its optical, electronic, and phononic properties [29–31]. In view of this property, we anticipate novel functions of black phosphorus, such as optical polarization sensors, anisotropic plasmonic devices, and thermoelectric applications [32, 33]. In the near future, however, we expect that applications requiring lower quality will come of age earlier, as in the case of graphene. One such application could utilize the broadband saturable absorption properties of black phosphorus suspensions [34].

Graphene, in conjunction with other 2D materials, offers a myriad of possibilities. As new players enter the field, we look forward to rapid progress from research to actual realization and commercialization of graphene-based applications.

References

1. Kim, K., Choi, J.-Y., Kim, T., Cho, S.-H., and Chung, H.-J. (2011). A role for graphene in silicon-based semiconductor devices, *Nature*, **479**, pp. 338–344.
2. Novoselov, K. S., Falko, V. I., Colombo, L., Gellert, P. R., Schwab, M. G., and Kim, K. (2012). A roadmap for graphene, *Nature*, **490**, pp. 192–200.
3. Chen, J. J., Wang, Q., Meng, J., Ke, X., Tendeloo, G. V., Bie, Y. Q., Liu, J., Liu, K., Liao, Z. M., Sun, D., and Yu, D. (2015). Photovoltaic effect and evidence of carrier multiplication in graphene vertical homojunctions with asymmetrical metal contacts, *ACS Nano*, **9**, pp. 8851–8858.
4. Koppens, F. H. L., Mueller, T., Avouris, P., Ferrari, A. C., Vitiello, M. S., and Polini, M. (2014). Photodetectors based on graphene, other two-dimensional materials and hybrid systems, *Nat Nanotechnol*, **9**, pp. 780–793.
5. Christensen, T., Yan, W., Jauho, A. P., Wubs, M., and Mortensen, N. A. (2015). Kerr nonlinearity and plasmonic bistability in graphene nanoribbons, *Phys Rev B Condens Matter Mater Phys*, **92**, pp. 121407.
6. Jadidi, M. M., Sushkov, A. B., Myers-Ward, R. L., Boyd, A. K., Daniels, K. M., Gaskill, D. K., Fuhrer, M. S., Drew, H. D., and Murphy, T. E. (2015). Tunable terahertz hybrid metal-graphene plasmons, *Nano Lett*, **15**, pp. 7099–7104.
7. Grigorenko, A. N., Polini, M., and Novoselov, K. S. (2012). Graphene plasmonics, *Nat Photon*, **6**, pp. 749–758.
8. Dean, C. R., Young, A. F., Meric, I., Lee, C., Wang, L., Sorgenfrei, S., Watanabe, K., Taniguchi, T., Kim, P., Shepard, K. L., and Hone, J. (2010). Boron nitride substrates for high-quality graphene electronics, *Nat Nanotechnol*, **5**, pp. 722–726.

9. Pacil, D., Meyer, J. C., Girit, Ç., and Zettl, A. (2008). The two-dimensional phase of boron nitride: Few-atomic-layer sheets and suspended membranes, *Appl Phys Lett*, **92**, pp. 133107.

10. Osada, M., and Sasaki, T. (2012). Two-dimensional dielectric nanosheets: Novel nanoelectronics from nanocrystal building blocks, *Adv Mater*, **24**, pp. 210–228.

11. Wilson, J. A., and Yoffe, A. D. (1969). The transition metal dichalcogenides discussion and interpretation of the observed optical, electrical and structural properties, *Adv Phys*, **18**, pp. 193–335.

12. Mattheiss, L. F. (1973). Band structures of transition-metal-dichalcogenide layer compounds, *Phys Rev B*, **8**, pp. 3719–3740.

13. Gupta, A., Sakthivel, T., and Seal, S. (2015). Recent development in 2D materials beyond graphene, *Prog Mater Sci*, **73**, pp. 44–126.

14. Wang, Q. H., Kalantar-Zadeh, K., Kis, A., Coleman, J. N., and Strano, M. S. (2012). Electronics and optoelectronics of two-dimensional transition metal dichalcogenides, *Nat Nanotechnol*, **7**, pp. 699–712.

15. Butler, S. Z., Hollen, S. M., Cao, L., Cui, Y., Gupta, J. A., Gutiérrez, H. R., Heinz, T. F., Hong, S. S., Huang, J., Ismach, A. F., Johnston-Halperin, E., Kuno, M., Plashnitsa, V. V., Robinson, R. D., Ruoff, R. S., Salahuddin, S., Shan, J., Shi, L., Spencer, M. G., Terrones, M., Windl, W., and Goldberger, J. E. (2013). Progress, challenges, and opportunities in two-dimensional materials beyond graphene, *ACS Nano*, **7**, pp. 2898–2926.

16. Zhao, H., Guo, Q., Xia, F., and Wang, H. (2014). Two-dimensional materials for nanophotonics application, *Nanophotonics*, **4**, pp. 128–142.

17. Xia, F., Wang, H., Xiao, D., Dubey, M., and Ramasubramaniam, A. (2014). Two-dimensional material nanophotonics, *Nat Photon*, **8**, pp. 899–907.

18. Zeng, H., Dai, J., Yao, W., Xiao, D., and Cui, X. (2012). Valley polarization in MoS_2 monolayers by optical pumping, *Nat Nanotechnol*, **7**, pp. 490–493.

19. Mak, K. F., He, K., Shan, J., and Heinz, T. F. (2012). Control of valley polarization in monolayer MoS_2 by optical helicity, *Nat Nanotechnol*, **7**, pp. 494–498.

20. Cao, T., Wang, G., Han, W., Ye, H., Zhu, C., Shi, J., Niu, Q., Tan, P., Wang, E., Liu, B., and Feng, J. (2012). Valley-selective circular dichroism of monolayer molybdenum disulphide, *Nat Commun*, **3:887**, pp. 1882.

21. Zhang, Y. J., Oka, T., Suzuki, R., Ye, J. T., and Iwasa, Y. (2014). Electrically switchable chiral light-emitting transistor, *Science*, **344**, pp. 725–728.

22. Ross, J. S., Klement, P., Jones, A. M., Ghimire, N. J., Yan, J., Mandrus, D. G., Taniguchi, T., Watanabe, K., Kitamura, K., Yao, W., Cobden, D. H., and Xu, X. (2014). Electrically tunable excitonic light-emitting diodes based on monolayer WSe_2 p-n junctions, *Nat Nanotechnol*, **9**, pp. 268–272.

23. Baugher, B. W. H., Churchill, H. O. H., Yang, Y., and Jarillo-Herrero, P. (2014). Optoelectronic devices based on electrically tunable p-n diodes in a monolayer dichalcogenide, *Nat Nanotechnol*, **9**, pp. 262–267.

24. Mak, K. F., He, K., Lee, C., Lee, G. H., Hone, J., Heinz, T. F., and Shan, J. (2013). Tightly bound trions in monolayer MoS_2, *Nat Mater*, **12**, pp. 207–211.

25. Ross, J. S., Wu, S., Yu, H., Ghimire, N. J., Jones, A. M., Aivazian, G., Yan, J., Mandrus, D. G., Xiao, D., Yao, W., and Xu, X. (2013). Electrical control of neutral and charged excitons in a monolayer semiconductor, *Nat Commun*, **4:1474**, pp. 2498.

26. Britnell, L., Ribeiro, R. M., Eckmann, A., Jalil, R., Belle, B. D., Mishchenko, A., Kim, Y. J., Gorbachev, R. V., Georgiou, T., Morozov, S. V., Grigorenko, A. N., Geim, A. K., Casiraghi, C., Castro Neto, A. H., and Novoselov, K. S. (2013). Strong light–matter interactions in heterostructures of atomically thin films, *Science*, **340**, pp. 1311–1314.

27. Yu, W. J., Liu, Y., Zhou, H., Yin, A., Li, Z., Huang, Y., and Duan, X. (2013). Highly efficient gate-tunable photocurrent generation in vertical heterostructures of layered materials, *Nat Nanotechnol*, **8**, pp. 952–958.

28. Morita, A. (1986). Semiconducting black phosphorus, *Appl Phys A Solids Surfaces*, **39**, pp. 227–242.

29. Tran, V., Soklaski, R., Liang, Y., and Yang, L. (2014). Layer-controlled band gap and anisotropic excitons in few-layer black phosphorus, *Phys Rev B Condens Matter Mater Phys*, **89**, pp. 235319.

30. Xia, F., Wang, H., and Jia, Y. (2014). Rediscovering black phosphorus as an anisotropic layered material for optoelectronics and electronics, *Nat Commun*, **5:4458**, pp. 5458.

31. Qiao, J., Kong, X., Hu, Z. X., Yang, F., and Ji, W. (2014). High-mobility transport anisotropy and linear dichroism in few-layer black phosphorus, *Nat Commun*, **5:4475**, pp. 5475.

32. Low, T., Roldán, R., Wang, H., Xia, F., Avouris, P., Moreno, L. M., and Guinea, F. (2014). Plasmons and screening in monolayer and multilayer black phosphorus, *Phys Rev Lett*, **113**, pp. 106802.

33. Fei, R., Faghaninia, A., Soklaski, R., Yan, J. A., Lo, C., and Yang, L. (2014). Enhanced thermoelectric efficiency via orthogonal electrical and thermal conductances in phosphorene, *Nano Lett*, **14**, pp. 6393–6399.

34. Wang, Y., Huang, G., Mu, H., Lin, S., Chen, J., Xiao, S., Bao, Q., and He, J. (2015). Ultrafast recovery time and broadband saturable absorption properties of black phosphorus suspension, *Appl Phys Lett*, **107**, pp. 091905.

Index

absorption 17, 28, 71, 73, 121, 122, 133, 134
  excited-state 133, 134, 139, 142
  multiphoton 133, 135, 136, 139
  photon 29, 53
  saturable 4, 20, 21, 37, 134
application 17, 20, 49, 50, 83, 85, 102, 104, 122–124, 143, 144, 148, 164, 166, 177, 179, 181, 182
  advanced 171
  commercial 131, 166
  industrial 14
  optical 142
  photonic 179
  thermoelectric 181
  touchscreen 167

bandgap 4, 66, 74, 90, 94, 157, 181
  electrical 8
  intrinsic 65
  moderate 74, 76
  near-zero 94
  zero 18, 74
band 18, 28, 29, 37, 61, 114
beam 57, 118
  dangerous 131
  infrared 107
  laser 51
bending angle 88, 89
BN *see* boron nitride
boron nitride (BN) 10, 123, 180
buffered oxide etch 12

capacitance 170–172
carbon nanotube (CNT) 134–136, 142, 143, 164–166
carrier 4, 69, 73, 104, 122, 152
  photo-exited 69
  photo-generated 94
cavity 30–32, 71, 72
  optical 70, 71
  resonant 71
cellular phone 164
chemical vapor deposition (CVD) 6, 10, 11, 60, 82, 88, 148, 151
CMOS *see* complementary metal-oxide semiconductor
CNT *see* carbon nanotube
complementary metal-oxide semiconductor (CMOS) 74
conductivity 16, 83, 87, 88
  electric 148
  electronic 16
  in-plane 167
  thermal 4, 82
contact resistance 53, 170, 179
coupling 59, 69, 72, 109, 110, 123
  antisymmetric 122
  critical 47, 50
  disk–disk 116
  electronic 95
  light–plasmons 115
  polarization-selective 19
  symmetric 122
CSL *see* current spreading layer
current density 83, 84, 86, 87, 89, 106

current spreading layer (CSL) 158
CVD *see* chemical vapor deposition

damping 112, 115, 119, 121
defect 8, 10, 16, 37, 82, 108, 117, 118
device adhesive 165
device architecture 8
device fabrication 8, 10
device 10, 42–44, 47, 50–53, 66, 67, 71, 72, 75, 87–90, 92, 93, 95, 149, 151, 154–158, 165, 166, 179, 180
  anisotropic plasmonic 181
  as-fabricated 70
  bilayer photovoltaic 88
  commercial 85
  graphene-based 11, 181
  graphene-based OLED 149
  graphene–silicon microring 50
  high-performance OPV 92
  light-emitting 177
  mode-locked 30
  nano-optoelectronic 118
  optical switch 76
  visual display 163
Dirac fermion 18, 104, 106, 123, 124
Dirac plasmon 117
Dirac point 17, 20, 52
dispersion 4, 14, 31, 32, 103, 109, 110, 114, 137, 143
doped graphene 18, 95, 105, 106, 114, 167
doping 95, 96, 104
  chemical 18, 61, 110
  electrical 122
  intrinsic 10
  nitrogen 83
  wet-chemical 13
DSSC *see* dye-sensitized solar cell
dye-sensitized solar cell (DSSC) 81–87

EELS *see* electron energy-loss spectroscopy
electric field 19, 20, 27, 42, 43, 48, 66, 94, 119, 181
electrode 51, 53, 66, 82, 84–88, 90, 148, 153–155, 171, 173
  anodic 89, 148
  back-gate 75
  current-spreading 155
  nanocrystalline 83
  sensor 170, 171
  transparent cathode 150
electron acceptor 88, 92, 93, 95
electron energy-loss spectroscopy (EELS) 107, 112, 113
electron gas 103, 105, 123
energy 33, 42, 44, 47, 103, 107, 116, 122, 134, 137
  electrical 147
  green 81
  half excitation 61
  high-pulse 27
  optical 69
  photon 17, 119
  solar 81
  thermal 134
epitaxial graphene 8, 108, 119
etching 6, 12, 13
  copper 13
  wet 73
extinction ratio 43, 47, 50, 58, 61, 62

fabrication 10, 11, 16, 43, 47, 62, 92
Fabry–Perot microcavity 71, 72
Fermi level 18, 42, 43, 47, 49, 52
Fermi velocity 105
FGS *see* functionalized graphene sheets
fiber 29, 31, 33, 36, 58, 59
fiber laser 29, 31–34, 36

film 6, 8, 59, 87, 90, 119, 148, 165, 168
  conductive 173
  damaged 165
  gold 32
  thin 16, 68, 90, 92, 103, 148, 157
  thin h-BN 124
  thin metal 57–59
flakes 14, 115
  graphene 14
  micrometer-sized 6
  thin 3
fluorine tin oxide (FTO) 82, 150
footprint 41, 42, 47
FTO *see* fluorine tin oxide
fullerenes 2, 134, 136, 143
functionalized graphene 20, 83, 87, 88, 90, 92, 148
functionalized graphene sheets (FGS) 85, 86

gate 48, 50, 52, 67, 114, 116
gate voltage 52, 70, 108, 112, 121
gating 104, 118
  electrical 18, 42, 110, 116
GQD *see* graphene quantum dot
graphene 1–8, 10–14, 17–21, 28–37, 49–53, 69–77, 81–85, 87–90, 92–96, 104–113, 115–124, 136–140, 142–144, 148–155, 177–182
  bilayer 10, 20, 109, 110
  double-layer 43, 47
  gapless 181
  gated 107
  multilayer 7, 9, 52, 121
  nitrogen-doped 83
  oxidized 33
graphene-based photodetector 66, 68, 70, 72, 74, 76, 77
graphene electrode 82–85, 87, 90, 153–156, 170, 171
graphene film 12–14, 30, 60, 61, 83, 84, 87, 89, 94, 144, 149, 153, 167, 173
graphene nanoribbon 112, 120, 182
graphene oxide 14, 17, 82, 83, 88, 137, 140, 142, 148, 155, 157, 158
graphene paper 87
graphene plasmon 104, 105, 110, 114, 115, 117–119, 124
graphene quantum dot (GQD) 90
graphene ribbon (GR) 112, 116, 120, 122
graphene sheet (GS) 2, 6, 69, 71–73, 82, 83, 90–92, 95, 96, 107, 108, 121, 137, 154, 157
graphite 2, 3, 6–8, 14, 15, 82, 102, 113, 137, 142, 148
GR *see* graphene ribbon
GS *see* graphene sheet

heterostructure 10, 37, 74, 76, 77, 119, 123, 179
highly oriented pyrolytic graphite (HOPG) 7
HIL *see* hole injection layer
HOF *see* hollow optical fiber
hole injection layer (HIL) 149, 150
hole injection process 149, 150
hole 29, 73, 75, 94, 95, 181
hole transport layer (HTL) 82, 90, 93, 149, 150
hollow fiber absorber 31
hollow optical fiber (HOF) 30
HOPG *see* highly oriented pyrolytic graphite
HTL *see* hole transport layer
hybrid solar cell 83, 93
hybrid system 132, 179, 182

indium tin oxide (ITO) 17, 82, 87–89, 148, 149, 158, 164, 165, 177
interaction 6, 7, 19, 43, 52, 61, 72, 107, 124, 134, 142, 164
  disk–disk 115
  electron–phonon 108
  electron–plasmon 114
  many-body 110, 121
  nonlinear 30
  optical near-field 118
  pairwise 105
  spin–orbit 124
interband transition 18, 19, 28, 44, 52, 106
interface 10, 59, 66, 94, 96, 102, 103, 181
intraband transition 18, 19, 44, 52
iPhone 167, 172
ITO *see* indium tin oxide
ITO device 88, 89

laser cavity 27, 29, 30, 32–34
laser 5, 28, 30–37, 50, 132, 136, 144, 179
layer 3, 6, 17, 34, 37, 44, 48, 60, 121, 150, 152, 158, 167–169, 172
  active 90, 92, 152, 155
  adhesive 13
  atomic 66
  copper 13
  current spreading 158
  dielectric 75
  electric node 172
  hybrid 181
  organic 149, 150
  thin metal 158
LCD *see* liquid crystal display
LED *see* light-emitting diode
light-emitting diode (LED) 5, 147, 148, 152, 154, 158, 181
liquid crystal display (LCD) 164, 167

massless Dirac fermion (MDF) 104, 106
MDF *see* massless Dirac fermion
metal contact 11, 66, 71, 157, 179, 182
metal organic vapor phase epitaxy (MOVPE) 156
microbubble 133, 134
microplasma 133, 134
micro-ribbon 52, 115
ML *see* mode-locking
mode-locked laser 31–33, 35
mode-locked pulse 30, 32–34
mode-locking (ML) 27, 35, 109
modulation depth 28, 33, 37, 42–44, 49, 50, 52, 53
modulator 5, 42–44, 47, 49, 50, 52, 53, 115, 121, 179
MOVPE *see* metal organic vapor phase epitaxy

NDG *see* nitrogen-doped graphene
nitrogen-doped graphene (NDG) 83
N-methyl-pyrrolidone (NMP) 14, 15, 137, 138
NMP *see* N-methyl-pyrrolidone
nonlinear scattering 133, 134, 136, 137

OLED *see* organic light-emitting diode
optical absorption 42, 43, 70–72, 122, 154
optical field 19, 20, 36, 70, 72, 118
optical interconnects 49, 50, 53

optical limiters 5, 131, 132, 134, 136, 138, 140, 142, 143
optical modulator 20, 41–43, 47–51
optical transition 18, 29, 42
optoelectronic device 5, 96, 151, 166
organic light-emitting diode (OLED) 149, 150, 164, 179
oscillation 59, 101–103, 107, 112

PCE *see* power conversion efficiency
phonon 19, 103, 107, 108, 110, 112, 114
photocarrier 66, 73, 76
photocurrent 66, 69, 71, 76
photodetector 5, 48, 66, 69, 73–75, 177, 179, 181, 182
photoluminescence 93, 157
photon 20, 29, 52, 103, 107, 110, 133
photovoltaic device 5, 104
Plank's constant 106
plasmon excitation 102, 103, 115, 116
plasmon 101–106, 108, 110, 112, 113, 115, 121–124
polarizer 5, 57, 62, 121, 179
polymer 6, 13, 62, 69, 90, 142, 165, 166, 172
power conversion efficiency (PCE) 83, 85, 88, 90, 93, 95
pulse 30, 32–34, 36, 134, 139, 140
pulse width 31, 33, 34, 36, 37

QCSE *see* quantum-confined Stark effect
QDSC *see* quantum dot solar cell
quantum-confined Stark effect (QCSE) 42
quantum dot solar cell (QDSC) 83

random-phase approximation (RPA) 18, 106, 107
reduced graphene 15, 16, 151, 155
reflector 50–52
refractive index 20, 42, 58, 59, 61, 143
resistance 70, 84, 85, 147, 173
resonance 51, 110, 111, 122, 124, 179
ribbon 110, 115, 116, 120, 122
RPA *see* random-phase approximation

SA *see* saturable absorption
saturable absorption (SA) 4, 20, 21, 31, 32, 37, 133, 134, 182
scattering 28, 112, 119, 135, 138
  surface-enhanced Raman 102
screen 163, 167, 172
semiconductor 3, 8, 41, 65, 94, 95
  complementary metal-oxide 74
  inorganic 95, 154
  low-bandgap 74
silicon waveguide 44, 48, 73
single-layer graphene 11, 28, 42, 47, 50, 52, 109, 110, 114, 121, 167
solar cell device 87, 88, 90, 93
solar cell 17, 81, 82, 84, 86–88, 90–92, 94–96, 177, 179, 181
  crystalline silicon 81
  graphene-based 95, 96
  low-cost plastic 90
  organic 81, 83, 89
  quantum dot 83
  Schottky junction 94
  semi-transparent 90
  thin-film 81, 95
solid state laser 33, 35–37

SPPs *see* surface plasmon polariton
SP *see* surface plasmon
strain 142, 154, 173–175
substrate 6, 8, 9, 11, 12, 14, 51, 52, 73, 82, 88, 89, 110, 114, 117, 119, 120, 148, 151, 168, 169
  low-temperature 85
  metal 11
  metal-catalyst-free 151
  polar 108, 109
  sapphire 157
surface plasmon polariton (SPP) 102–104, 107, 108, 122
surface plasmons (SP) 19, 59, 103, 104, 107, 110–113, 115, 117, 119, 121, 122

TCF *see* transparent conductive film
TCO *see* transparent conducting oxide
touch 163, 167, 172
touch pressure 170
touchscreen 5, 17, 163–166, 171, 172, 174, 175, 177, 179
touch sensor 170
touch signal 167
toxicity 16, 165
transition 18, 47, 61, 74, 110, 133
transmittance 4, 17, 82, 111, 123, 131, 132, 135, 136, 139–141, 158, 167, 174
transparent conducting oxide (TCO) 154, 165
transparent conductive film (TCF) 6, 164, 168
transparent electrode 75, 82, 83, 85, 150, 164–167, 170

waveguide 47, 49, 59, 60, 73, 115, 122
  optical 72, 73
  planarized 49
  ring 48
  silicon nitride 50
wave vector 103, 104, 107, 112, 122
WF *see* work function
work function (WF) 77, 90, 95, 149, 154–156, 158